J. A Blair

The organic Analysis of potable Waters

J. A Blair

The organic Analysis of potable Waters

ISBN/EAN: 9783743330047

Manufactured in Europe, USA, Canada, Australia, Japa

Cover: Foto ©berggeist007 / pixelio.de

Manufactured and distributed by brebook publishing software
(www.brebook.com)

J. A Blair

The organic Analysis of potable Waters

THE ORGANIC ANALYSIS

OF

POTABLE WATERS

BY

J. A. BLAIR, M.B., C.M., D.Sc.Edin.
L.R.C.P.Lond.

LONDON
J. & A. CHURCHILL
11, NEW BURLINGTON STREET
1890

CONTENTS

PREFACE

THE greater part of the experimental work contained in the following pages was done in the Laboratory of the Medical Jurisprudence and Public Health Department of the University of Edinburgh. I gratefully acknowledge the kindly interest shown, and encouragement given by Sir Douglas Maclagan whilst I was engaged with the work.

It was Dr Hunter Stewart, of the Public Health Department, who suggested to me the possibility of proving that Kjeldahl's Process for estimating Organic Nitrogen can be applied to potable waters; and he also pointed out to me the possible utility of a series of analyses of potable waters made by the "Oxygen and Albuminoid Ammonia Processes." I gratefully acknowledge not only these suggestions of his, but also the information given by him, from time to time, on subjects bearing on the work in general, and the references given by him to the literature of the subject in this, and other countries.

J. A. BLAIR.

MAIDSTONE;
July, 1890.

THE
ORGANIC ANALYSIS OF POTABLE WATERS

INTRODUCTION

THIS branch of analytical work is handicapped by two great drawbacks, namely, the very small quantity of organic matter which has to be dealt with, and the impossibility, in the present state of chemical science, of measuring these organic bodies as such. Much of the organic matter in potable waters is highly complex in composition, unstable, and readily induced to undergo chemical changes, if the conditions of its existence are modified. In order to meet these difficulties, analysts have resorted to analytical methods, whereby these highly complex bodies are broken up into more elementary, and more stable compounds. Nitrogen is a constituent of many organic compounds —more especially of albuminoid bodies,—and, therefore, has been used as a partial indicator of the

1

quantity of organic matter contained in a sample of water. Carbon is also present in many organic bodies, and is also used as a partial indicator of the quantity of organic matter. Speaking generally, organic bodies whilst decomposing do not take up either nitrogen or carbon, therefore analysts prefer to measure the quantity of these elements, rather than the quantity of hydrogen and oxygen, which also are contained in very many organic compounds. Many organic compounds undergoing decomposition absorb oxygen, and the quantity of oxygen thus absorbed may be used to measure the relative quantity of organic matter present in the water examined.

The very small amount of organic matter, in a reasonable quantity of average potable waters, has forced chemists to resort to very delicate processes of analysis. This delicacy of operation tends to make the average experimental error considerable, even in trained hands. The experimental error is least with the analytical process which assures accuracy of measurement, consistent with simplicity of apparatus and the minimum of precautions.

In this country, the organic impurity of potable waters is usually estimated by one of three methods, or by a combination of them, namely:

(1) The Albuminoid Ammonia Process—brought out by Messrs. Wanklyn, Chapman, and Smith (1867).

(2) The Estimation of Organic Carbon and Nitro-

gen—brought out by Drs Frankland and Armstrong (1868).

(3) The Oxygen Process—as elaborated by Prof. Tidy (1879).

The various analytical processes will be considered in the following order :

I. The Albuminoid Ammonia and Oxygen Processes, and then some considerations in reference to the analyses of waters by both methods.

II. The Oxygen Process at 100° C.

III. The Estimation of the Organic Nitrogen by the Sulphuric Acid Process.

IV. The Estimation of the Organic Carbon by the Sulphuric Acid and Permanganate Process.

PART I

THE ESTIMATION OF THE ORGANIC IMPURITY OF POTABLE WATERS BY THE "ALBUMINOID AMMONIA PROCESS," AND BY THE "OXYGEN PROCESS AT ORDINARY TEMPERATURES"

THESE two processes are similar in only giving partial results. The " Oxygen Process," as usually carried out, does not completely oxidise all the organic matter; and the "Albuminoid Ammonia Process" does not convert all the organic nitrogen, which may be present in a water, into ammonia. These partial measurements of oxygen required, and nitrogen present are in some respects unsatisfactory. Putting aside the question of micro-organisms and specific poisons, sanitarians seem to agree that the danger to health, from drinking a water containing organic matter, does not depend so much upon the quantity of organic matter contained in the water, as upon the capability of the matter for readily undergoing chemical changes. The albuminoid ammonia and oxygen processes, to some extent, measure the potentiality of the organic matter for rapid chemical changes.

Both methods take advantage of the fact that organic matter, in the presence of suitable oxidising agents, readily takes up oxygen, and becomes partially oxidised into less complex, but more chemically stable, compounds. In the oxygen process, the quantity of oxygen used up in this way is made the indicator for judging of the organic impurity of the sample of water; but in Mr Wanklyn's process, the ammonia evolved by the combined action of caustic alkali and an oxidising agent is the indicator of the amount of organic impurity. It is claimed for both processes that the results obtained are sufficiently uniform to enable the analyst to form an opinion of the fitness, or unfitness of the water for potable purposes.

The Albuminoid Ammonia Process may be carried out in the following way.

Certain solutions are required, and instructions for their preparation will be found in Appendix A.

The solutions are—

(1) A dilute solution of ammonium chloride of the strength that each c.c. contains $\frac{1}{100}$ milligramme of ammonia.

(2) Nessler's reagent.

(3) A strongly alkaline solution of permanganate of potash.

(4) A saturated solution of sodium carbonate.

Take a 32 oz. distillation flask, with a side tube in the neck, and cleanse it well with good tap water; if not previously used for this purpose, wash out the flask with strong sulphuric acid, and again with tap water till the drippings are neutral to litmus paper. Fit the flask with a good, well-fitting cork covered with a layer of tin-foil, and attach the distillation flask to a clean Liebig condenser, in the usual way, packing with tin-foil or writing-paper, if necessary, and supporting the flask on wire-gauze over the ring of an iron stand. It is very necessary to have all measuring flasks and apparatus scrupulously clean. Measure out half a litre of the water to be analysed; pour it into the flask through a long funnel. If the water is acid, add 10 c.c. of the sodium carbonate solution. Boil the water briskly, and collect the distillate in four stoppered flasks of 50 c.c. capacity. The first 50 c.c. of the distillate is now poured into a nesslerising glass, and 2 c.c. of Nessler's reagent added to it; mix well by gently shaking the glass, and, if ammonia is present, it will strike a yellow colour. Into another nesslerising glass run in from a burette as much of the weak solution of ammonium chloride as you think equal to the ammonia in the distillate, make up to 50 c.c. with distilled water—free of ammonia—and add 2 c.c. of Nessler's reagent. If sufficient ammonium chloride has not been added to give the same colour as in the 50 c.c. of distillate, run in rapidly successive quantities of ammonium chloride solution till the colours correspond; when the proper quantity of solution required has been thus found, repeat the operation in a clean nesslerising glass. When nesslerising, stand the glasses on a white tile. The "Free Ammonia" is estimated in this way, and Mr Wanklyn recommends adding to the total quantity one third of the quantity found, and in this way avoiding the necessity of nesslerising the remaining 150 c.c. of distillate.

The Albuminoid Ammonia.—Take 50 c.c. of the alkaline permanganate solution, boil it in a small flask for five minutes, in order to free it of traces of ammonia, and then add it to the 300 c.c. of the water left in the distillation flask. Resume the boiling process, and collect the distillate in successive quantities of 50 c.c.; nesslerise each of these 50 c.c. in the above manner till the last 50 c.c. contain no ammonia, then stop any further distillation.

The free and albuminoid ammonia found in a half-litre of water, if doubled will give the ammonia as parts per million, or milligrammes per litre.

The Oxygen Process at Ordinary Temperatures.

Solutions required (for preparation see Appendix B) :

(1) Solution of potassium permanganate—10 c.c. of the solution are capable of yielding up one milligramme of oxygen.

(2) Dilute sulphuric acid.

(3) Solution of potassium iodide.

(4) Solution of sodium hyposulphite.

(5) Solution of starch.

Clean out with strong sulphuric acid four flasks of about 20 oz. capacity, and then rinse out with good tap water. Into each of two of these flasks pour 250 c.c. of the water to be examined, and into the other two flasks pour 250 c.c. of distilled water; to each flask add 10 c.c. of the dilute sulphuric acid and 10 c.c. of the permanganate solution ; cork the flasks with clean corks or cover with paper, and place them at a window exposed to a good light. Note the time when the permanganate is added to the waters, and if the pink colour of the water become very pale before the allotted time is up, it is advisable to add another 10 c.c. of permanganate solution. At the end of a quarter of an hour, remove one flask containing the sample of water and one containing distilled water, and add to each flask five drops of the solution of potassium iodide for each 10 c.c. of permanganate added. Rotate the flasks gently till the colour is a clear, brilliant yellow. From a burette run into the flask containing distilled water the hyposulphite solution till the yellow colour is very faint; now add sufficient starch solution to produce a deep blue colour; continue to run in slowly the hyposulphite solution till the blue colour just disappears, and note the quantity of hyposulphite solution used. Treat the sample of water in exactly the same way. At the end of three hours, treat the contents of the remaining two flasks in the same way.

Calculation required.—In all cases, the quantity of hyposulphite needed to decolorise the iodine is proportionate to the quantity of permanganate in the water, or, in other words, to the quantity of oxygen available for oxidising purposes. In the blank experiment with distilled water, the quantity of hyposulphite used represents one milligramme of oxygen, or 10 c.c. of permanganate solution. If the number of c.c. of hyposulphite solution used by the sample of water is deducted from the number of c.c. of hyposulphite solution used by the distilled water, the difference will represent the oxygen used up by the sample of water.

The equation is stated thus:—c.c. of hyposulphite used by distilled water : c.c. of difference :: one milligramme of oxygen : answer

$$= \frac{\text{c.c. of difference}}{\text{c.c. of blank exp.}} = \begin{array}{l}\text{oxygen used expressed in the}\\\text{fraction of a milligramme.}\end{array}$$

The chemical equations, representing the reactions which take place at the various stages of the process, are given in Appendix B.

Sulphuretted hydrogen, ferrous salts, and nitrites will reduce permanganate. In the case of waters containing any one of these chemical compounds, in order to prevent errors arising in the estimation of the oxygen used by the organic matter, Prof. Tidy recommends heating the acidified water to get rid of the sulphuretted hydrogen, and the estimation of the oxygen used up in five minutes when ferrous salts, or nitrites are present. The oxygen thus used is deducted from the amount of oxygen used up in a quarter of an hour and in three hours, in order to find the quantity used up by the organic matter. Divide the oxygen used by 2·5, and the answer gives the oxygen as parts per 100,000 of the water.

The following tables give a series of analyses of various organic solutions and waters by the albuminoid ammonia and oxygen processes.

TABLE I

Animal Organic Solutions.

Solution I of fresh White of Egg, made by dissolving 2·518 grammes of the fresh material, and 1·10 grammes of freshly ignited sodium carbonate in 200 c.c. of distilled water. Solution was kept in a stoppered flask.

Solution II of fresh White of Egg, made by dissolving 0·525 gramme of the fresh material, and 0·50 gramme of sodium carbonate in 100 c.c. of distilled water. The solution was kept in a stoppered flask.

Solution III of fresh White of Egg, made by dissolving 1·388 grammes of the fresh material, and 1·39 grammes of freshly ignited sodium carbonate in 100 c.c. of distilled water.

Ammonia and oxygen are stated as parts per million.

No. of solution.	Age of solution.	Quantity used.	Oxygen used in		Ammonia yielded.
			¼ hour.	3 hours.	
I	6 days	0·7 c.c.	100·0	178·8	158·5
II	3 „	4·0 „	33·9	47.2	49·0
II	5 „	3·0 „	34·4	48·4	48·0
II	5 „	2·0 „	34·6	47·3	42·0
III	1 day	1·0 „	77·6	103·2	180·0

TABLE II

Water polluted with Organic Animal Matter.

1·0 c.c. of Fresh Urine was added to 2 litres of Edinburgh City water.

Oxygen and albuminoid ammonia are stated as milligrammes per litre, or parts per million.

Water.	Age.	Quantity used.		Oxygen used in		Alb. ammonia yielded.
		Tidy.	Wanklyn.	¼ hour.	3 hours.	
Edinburgh City	Fresh	250 c.c.	500 c.c.	0·82	2·15	0·17
Same sample polluted as above	1 day old	50 ,,	20 ,,	1·29	2·95	5·20
	2 days old	50 ,,	20 ,,	1·13	2·57	3·55
	3 ,,	50 ,,	20 ,,	1·09	2·42	—
	5 ,,	50 ,,	20 ,,	0·98	2·51	3·47
	7 ,,	100 ,,	50 ,,	1·16	2·49	3·08
	12 ,,	100 ,,	50 ,,	0·94	2·52	3·96
1 part above sample plus 2 parts aq. dest.	14 ,,	200 ,,	100 ,,	0·38	0·72	0·68

Sewage—made by making a 1 per cent. solution of fresh urine in distilled water.

Oxygen and albuminoid ammonia stated as parts per million.

Water.	Age.	Quantity used.	Oxygen used in		Alb. ammonia yielded.
			¼ hour.	3 hours.	
Sewage.........	Fresh	1·0 c.c.	13·30	24·30	90·00

TABLE III

Effluent Waters.

No. 1. Sewage water taken from the main sewer as it enters Craigentinny Meadows. Heavy rains a few hours previously.

No. 2. Effluent water from Craigentinny Meadows, just before it runs into the sea.

No. 3. Edinburgh City water polluted with the above effluent water (No. 2).

Oxygen and albuminoid ammonia are stated as parts per million, or milligrammes per litre.

No. and age.	Oxygen used in			Alb. ammonia yielded.
	5 minutes.	¼ hour.	3 hours.	
No. I.—Fresh	9·49	13·91	26·08	13·52
„ 2 days	—	11·80	20·74	8·00
No. II.—3 days...............	8·22	10·30	18·99	5·42
Edinburgh water ...	—	—	2·02*	0·10
No. III.—Ditto plus 20 c.c. effluent II	—	—	2·34*	0·21

* The oxygen stated is the quantity left after deducting the quantity used up in five minutes from the total quantity used in three hours.

TABLE IV

Vegetable Solutions.

Strychnia I.—A solution made by dissolving 100 milligrammes of pure crystals of strychnia in sulphuric acid, and diluting the solution up to 100 c.c. with distilled water.

Strychnia II.—Same strength as No. 1 solution.

Brucia.—Solution made as above, and 1 c.c. contained 1 milligramme of brucia.

Muriate of Morphia.—Solution of same strength as above.

Tea.—Solution made by digesting 0·50 gramme of tea in 100 c.c. of cold distilled water. At the end of one hour the solution was filtered, and made up to 150 c.c. with distilled water.

Oxygen and ammonia are stated as parts per million.

Substance.	Quantity used.	Oxygen used in		Ammonia yielded.
		¼ hour.	3 hours.	
Strychnia I	0·5 c.c.	110·0	150·8	72·0
„ II	1·0 „	101·0	119·3	70·0
Brucia	3·0 „	265·0	489·0	58·0
Morph. mur.	1·0 „	377·0	432·0	35·0
„ carefully dried...	4·0 „	341·0	559·0	33·3
Tea................................	1·0 „	67·5	142·5	36·0
„ 12 days old	5·0 „	54·0	86·6 {	Alb. am. 10·6

Table V

Vegetable Solutions (continued).

Vegetable Albumen.—Solution made by digesting chips of raw potato in cold distilled water, and at the end of one hour decanting off the water.

Wheaten Flour.—Solution made by adding 100 milligrammes of the flour and 100 milligrammes of caustic soda to distilled water, and making the solution up to 100 c.c.

Soluble Albumens of Wheaten Flour.—One gramme of the same flour as above was digested in 100 c.c. of distilled water, containing a few drops of liquor potassæ, for eighteen hours, and the solution was then filtered.

Oxygen and ammonia are stated as parts per million.

Substance.	Quantity used.	Oxygen used in		Total ammonia yielded.
		¼ hour.	3 hours.	
Potato solution	10·0 c.c.	18·9	60·8	29·7
Flour albumens....................	10·0 „	43·9	118·4	123·9
Flour solution	10·0 „	5·8	10·5	12·0

TABLE VI

Natural Peaty Waters.

Oxygen and albuminoid ammonia are stated as parts per million, or milligrammes per litre.

Water.	Oxygen used in		Alb. ammonia yielded.
	¼ hour.	3 hours.	
Edinburgh water from reservoir on Alnwick Hill, not filtered	0·99	2·48	0·265
Loch Spalandar—Old peat	5·56	10·69	0·215
„ 2nd sample...	—	11·18	0·287
Patna Moor—Surface stream ...	6·01	12·28	0·400

TABLE VII

Natural Waters, filtered previous to Use.

Oxygen and albuminoid ammonia are stated as parts per million, or milligrammes per litre.

Date.	Water.	Oxygen used in		Alb. ammonia yielded.
		¼ hour.	3 hours.	
Dec. 15, 1887	City supply, Edinburgh	1·02	2·24	0·187
„ 16, „	„ „	0·88	2·20	0·183
Jan. 11, 1888	„ „	0·82	2·15	0·170
Feb. 8, „	„ „	—	2·02	0·100
Mar. 12, „	„ „	0·70	1·87	0·128
„ 13, „	„ „	0·74	1·90	0·127
Apr. 17, „	„ „	0·87	2·07	0·100
Nov. 22, „	„ „	0·76	2·18	0·164
Dec. 7, „	„ „	0·91	1·93	0·113
Jan. 14, 1889	„ „	0·86	1·66	0·098
Mar. 27, „	„ „	0·64	1·62	0·126
Apr. 4, „	„ „	0·64	1·46	0·132

TABLE VIII

Well Waters.

Oxygen and albuminoid ammonia are stated as parts per million, or milligrammes per litre.

Water.	Oxygen used in		Alb. ammonia yielded.
	¼ hour.	3 hours.	
St Anthony's Well, Queen's Park	0·14	·0·22	0·115
„ „ „	0·00	·0·00	0·014
St Margaret's Well, „	0·12	·0·29	0·127
„ „ „	0·04	·0·13	0·052
„ „ „	0·09	·0·26	0·077
Manse Well, Straiton	0·34	0·89	0·315
„ „ „	1·42	2·72	0·355
„ „ „	0·26	0·42	0·055
„ „ „	0·46	1·01	0·116
Surface Well, „	0·21	0·45	0·101
Toll Well, „	0·03	0·13	0·093
„ „ „	0·28	0·51	0·180
„ „ „	0·05	0·17	0·053
„ „ „	0·06	0·11	0·025
Village Well, No. I, Straiton	0·03	0·25	0·066
„ „ No. II, „	0·00	0·09	0·063
Public School Water „	0·00	0·12	0·024
Trough, Powburn........................	0·00	0·00	0·017

TABLE IX

Well Waters (continued).

Oxygen and albuminoid ammonia are stated as parts per million, or milligrammes per litre.

Water.	Oxygen used in		Alb. ammonia yielded.
	¼ hour.	3 hours.	
Manse Well, Kirkmichael.........	0·18	0·24	0·037
Village ,, ,, ...	0·07	0·20	0·090
No. I ,, ,, ...	0·06	0·12	0·037
No. II ,, ,, ...	0·15	0·30	0·110
No. III ,, ,, ...	0·00	0·25	0·060
No. IV ,, ,, ...	0·00	0·10	0·040
No. I ,, Patna	0·00	0·12	0·035
No. II ,, ,, ...	0·00	0·00	0·025
Peterhead Prison Supply	0·70	1·86	0·170
,, Sample II...............	0·92	1·80	0·152
Water No. I.—Peterhead.........	0·56	1·24	0·100*
,, II ,, ...	0·00	0·33	0·178*
,, III ,, ...	0·97	2·34	0·194
,, IV ,, ...	0·35	0·86	0·120
Spring.—The Braids...............	0·02	0·07	0·035
Spout.—Blackford Hill............	0·06	0·21	0·042
Well.—Northumberland Street, Edinburgh	0·18	0·44	0·128

* Much free ammonia, and in No. II a large quantity of ferrous chloride. The oxygen used up by the ferrous salts is deducted in the above table.

TABLE X

Polluted Waters.

Containing much organic matter of animal and vegetable origin.

Oxygen and albuminoid ammonia are stated as parts per million, or milligrammes per litre.

Water.	Oxygen used in		Alb. ammonia yielded.
	¼ hour.	3 hours.	
St Margaret's Loch, Queen's Pk.	0·66	1·39	0·700
„ „ „	0·69	1·56	0·485
„ „ „	0·94	1·96	1·247
„ „ „	0·88	1·43	0·380
Dunsappy „ „	0·58	1·44	0·550
„ „ „	0·80	1·59	0·245
Duddingston „ „	1·21	2·42	0·367
„ „ „	1·23	2·70	0·350
„ „ „	1·26	2·98	0·277
River Esk, Hawthornden.........	2·88	5·92	0·730
„ same sample	—	5·55	0·730
Water of Leith, Colinton Bridge	3·60	6·97	1·050
„ same sample ...	3·75	6·90	0·765
„ Dean Bridge ...	2·43	7·16	0·505
„ „ ...	2·29	6·92	0·466
„ Bonnington ...	5·44	18·90	2·720
„ „ ...	3·47	13·59	2·060
„ „ ...	3·89	15·13	2·280

The organic impurities of potable waters are divided, for practical purposes, into two divisions :

1. Organic substances of animal origin.
2. Organic substances of vegetable origin.

The first group of organic bodies is regarded as much more dangerous to health than the latter. Fluids containing organic bodies of animal origin often carry the germs of diseases associated with the presence of animals, and for this reason it is of great importance that, on analysis, the organic matters in a water can be assigned to their proper class. Unfortunately, when the quantity of organic matter in a sample of water is very small, it is not always possible to do this, and yet it is just in these cases that the question is so important. When the quantity of organic matter is large, the water is condemned irrespective of the origin of the matter.

The series of analyses, contained in the tables alluded to, was made with the object of showing the results got when examining the same organic solution by the albuminoid ammonia and oxygen processes. In order to facilitate the comparison between the two methods, the results in both cases are stated as parts per million. When very small quantities of the organic solutions were used, the alkaline permanganate was boiled in a distillation flask with distilled water till the distillate gave no colour reaction with

Nessler's reagent, and then the organic fluid was dropped into the permanganate solution.

Solutions of Organic Matter of Animal Origin.

Table I shows that fresh white of egg yields up as much ammonia as it uses up of oxygen in three hours, and that, in all cases, the oxygen used up in a quarter-hour is considerably less than the quantity of ammonia yielded.

Table II shows that fresh urine yields up considerably more ammonia than it uses up of oxygen in three hours, and the difference is all the more marked if compared with the oxygen used in a quarter-hour. The table also shows that these results hold good when urine is mixed with a natural water, and let stand for twelve days before analysing.

Table III shows that effluent waters from sewage farms, when not thoroughly purified, contain organic substances in a very readily oxidisable condition, and consequently the oxygen used up in three hours is, at least, twice as much as the albuminoid ammonia yielded—in some cases the ratio is even greater. The effluent waters examined doubtless contained organic substances of vegetable as well as of animal origin. The oxygen used in a quarter-hour is about equal to the albuminoid ammonia yielded. The ratio of oxygen to ammonia remained unchanged when the effluent water was added to a sample of the water supplied.to Edinburgh city.

Solutions of Organic Matter of Vegetable Origin.

Table IV shows that the vegetable alkaloids experimented with use up far more oxygen in three hours than ammonia yielded, the ratio of oxygen to ammonia being roughly as 2 to 1, and in some cases much greater. The oxygen used up in a quarter-hour, in all cases, considerably exceeded the ammonia yielded. The same remarks hold good in the case of tea.

Table V shows the experimental results obtained from vegetable albumens. The solution of albumen from potato resembles the effluent waters examined, in that the oxygen used in three hours has a ratio of 2 to 1 to the ammonia yielded. The albuminoid solutions from wheaten flour gave the same ratio of oxygen used to ammonia yielded as in the case of egg albumen.

Table VI shows the oxygen used in three hours by peaty waters is greatly in excess of the albuminoid ammonia yielded, the ratio ranging in the samples examined from 10 to 1 up to 50 to 1. The oxygen absorbed in a quarter-hour exceeds the ammonia yielded in a ratio ranging from 4 to 1 up to 25 to 1. The ratio seems to be highest in waters from old peat.

Table VII shows that when peaty waters are filtered the oxygen absorbed in three hours still exceeds the ammonia yielded in a ratio of 25 to 1.

Tables VIII and IX contain waters which are nearly all of great purity—in some five or six cases the waters are suspicious. In all cases, the oxygen absorbed in three hours exceeds the albuminoid ammonia yielded, except in cases where the water is of extraordinary organic purity. In most cases, the oxygen is in considerable excess over the ammonia, and in a few cases the ratio is as 9 to 1. The oxygen absorbed in a quarter-hour in some cases exceeds the albuminoid ammonia, and in other cases it is less.

Table X. In this series of polluted waters the oxygen absorbed in three hours exceeds the albuminoid ammonia in a ratio ranging from 1½ to 1 up to about 10 to 1. The oxygen absorbed in a quarter-hour in all cases, at least, equals the albuminoid ammonia, and generally it is greatly in excess of the ammonia.

Summary of Results obtained from the Tables.

1. Solutions of egg albumen and of urine yield as much as, or more albuminoid ammonia than they ab-

sorb of oxygen in three hours. This is also true of
the vegetable albumen of wheaten flour, but in the
case of the albumen from potato, the oxygen absorbed
in three hours is about twice as much as the albuminoid ammonia yielded.

II. Effluent waters contain the organic matter in a
highly oxidisable condition, and much of the albuminoid ammonia has been converted into free ammonia,
consequently the oxygen used up in three hours was
found to be one and a half times to twice as much as
the albuminoid ammonia, even when the oxygen used
by the nitrites is deducted from the total quantity used.
If the oxygen used in five minutes (due to nitrites) is
deducted from the oxygen absorbed in a quarter-
hour, it is found that the albuminoid ammonia exceeds
the oxygen used in a quarter-hour in a ratio of $2\frac{1}{2}$
to 1 up to $3\frac{1}{2}$ to 1.

III. Well waters containing only vegetable organic
matter, as a rule, absorb much more oxygen in three
hours than they yield of albuminoid ammonia. This
may be due to the waters containing oxidisable substances, which do not yield ammonia, such as starch,
gum, sugar, and vegetable alkaloids containing little
nitrogen. Complete oxidation of albumen requires
nearly ten times as much oxygen as it is capable of
yielding of ammonia.

IV. Vegetable alkaloids usually contain little nitrogen compared with the large quantity of carbon, con-

sequently the oxygen absorbed in three hours exceeds the ammonia yielded.

V. Polluted waters, containing both vegetable and animal organic matters. If Edinburgh sewage, as it enters Craigentinny Meadows, be taken as a type of sewage waters in general, it is evident that a natural water when polluted with sewage water will still absorb much more oxygen in three hours than it will yield up albuminoid ammonia, because sewage water itself has its oxygen absorbed in three hours in excess of the albuminoid ammonia at least in a ratio of $1\frac{1}{2}$ to 1. Of course such a polluted water will contain much "free ammonia." Edinburgh water, in which the oxygen absorbed in three hours exceeded the albuminoid ammonia yielded in the ratio of 20 to 1, after the addition of 20 c.c. of the Craigentinny effluent had this ratio reduced to about 11 to 1.

VI. Peaty waters use very large quantities of oxygen in three hours compared with the albuminoid ammonia yielded, the ratio of oxygen to albuminoid ammonia ranging from 10 to 1 up to 50 to 1, or even greater.

The following is Mr Wanklyn's classification of waters, based upon his albuminoid ammonia process.

A water is
{
  of great purity if the alb. ammonia is less than 0·05 part per million.

  organically safe if the alb. ammonia is less than 0·10 part per million.

  a dirty water if the alb. ammonia is more than 0·10 part per million.
}

Professor Tidy's classification of waters, based upon the oxygen used, is—

A water is
{
  of great organic purity if the oxygen used in three hours is less than 0·05 part per 100,000.

  of medium organic purity if the oxygen used in three hours is from 0·05 to 0·15 part per 100,000.

  of suspicious organic purity if the oxygen used in three hours is from 0·15 to 0·21 part per 100,000.
}

I believe that the range of suspicious waters is extended up to 0·30 part of oxygen per 100,000, and above this point waters are considered as unfit for domestic use.

The Advantages of examining Samples of Waters by both the Albuminoid Ammonia and Oxygen Processes.

If one process be trusted to only, errors of judgment may be made, and a water may be condemned as unfit for use when it is really a drinkable water. Of the well waters examined (see Tables VIII and IX) eight of them would be pronounced suspicious, or even dirty waters in some cases, if the albuminoid ammonia process had been alone trusted to; if these eight waters had been analysed only according to the oxygen process most of them would have been pronounced as waters of great purity, and all of them stated to be potable waters.

On the other hand, peaty waters are liable to be condemned as suspicious or dirty waters by the oxygen process, when pronounced potable according to Mr Wanklyn's classification.

It is evident that these two classifications of waters are not in harmony with one another. This state of affairs has led to contradictory reports being made upon the same water, where the analysts had relied on one process only for estimating the organic matter.

If a water is not condemned by much " free ammonia " or nitrites, and if the chlorine is moderate in quantity, the above experiments seem to indicate the following conclusions :

(1) If the oxygen absorbed in three hours does not exceed 0·05 part per 100,000 the quantity of albuminoid ammonia yielded is of little consequence as long as it does not exceed 0·20 part per million.

(2) When the oxygen absorbed in three hours is as high as 0·10 part per 100,000 the albuminoid ammonia should be at least ten times less, or not more than 0·12 part per million.

(3) If the oxygen absorbed in three hours rises as high as 0·15 part per 100,000 the albuminoid ammonia should be at least twelve times less, or not more than 0·12 part per million.

(4) When the oxygen absorbed in three hours is more than 0·15 part per 100,000 the albuminoid ammonia should be at least fifteen times less, or not more than 0·10 part per million, and running up to 0·20 part per million in proportion as the quantity of oxygen absorbed rises above 0·15 part per 100,000. Waters absorbing so much oxygen in three hours can only be safe if the organic matter is peaty, and in these cases the ratio of oxygen to ammonia is naturally high.

If a water fulfils these conditions it may be regarded as a potable water, as far as the organic matter is concerned.

PART II

THE OXYGEN PROCESS AT 100° C.

THE albuminoid ammonia and oxygen processes are faulty in several ways.

The chief objection to the oxygen process at ordinary temperatures is—

By the oxygen process, as usually carried out, the organic matter is only partially oxidised ; at best the oxygen absorbed in three hours is only about a quarter of the oxygen required for complete oxidation of the organic matter, and in many cases it is a much smaller fraction of the total quantity required. The albuminoid solution from wheaten flour, and the solution of moist white of egg only absorbed in three hours about a twenty-fifth of the oxygen required for the complete oxidation of the organic matter in the solutions, and, even under the most favorable conditions, the albuminoid solution from potato only absorbed in three hours a little more than one-fifth of the oxygen required for complete oxidation.   During three hours, at ordinary temperatures, starch and cane sugar absorb almost no oxygen compared with the quantity required for the complete oxidation of these substances.   The proof of these statements will be shown later on, in a series of tables.

After the consideration of this objection to the oxygen process as usually carried out, the question naturally arises—Can this objection to the oxygen process be overcome or minimised ?

Hitherto, the best method of carrying out the oxygen process is the one recommended by Professor Tidy, the working details of which were published in the ' Chemical Society's Journal' for 1879.   Since then his plan has usually been followed, some analysts modifying the time of exposure of the water to the action of the oxidising agents, and others modifying the temperature.   It has been suggested to keep the water in stoppered bottles at a temperature of 80° F. for four hours, whilst it is exposed to the action of the permanganate.   If the quantity of oxygen absorbed by a water in a given time can be used as a reliable basis for classifying the organic purity of waters, even when this quantity of oxygen is only a small fraction of the total quantity the organic matter is capable of absorbing, it is reasonable to suppose that the nearer the quantity absorbed can be made to approach the quantity required the more satisfactory will be the classification. The oxygen process is simply used as a means of roughly determining, in the absence of nitrites, ferrous salts, and sulphuretted hydrogen, the relative quantity of organic matter present in a potable water; but were it proved that the deoxidising property of organic matter is *per se* hurtful to the human body, then it is evident that the temperature of the human body would be the proper one at which to carry out the oxygen process. In the absence of this proof, the temperature of the water during the process of analysis is not of vital importance. Heat is a great promoter of chemical action, and the oxidation of the organic matter is much more complete if the mixture of water, acid, and permanganate is boiled for some time.

It is necessary to prove that the results obtained by boiling in this way can be relied upon.

In order to furnish this proof, the following points must be considered.

(1) Is a solution of permanganate of potash at 100° C. stable in the absence of reducing agents ?

(2) If stable at this temperature, is this stability affected by the addition of sulphuric acid or alkali to the solution ?

(3) Does the reducing power of organic solutions vary with the quantity of acid or alkali present in the solution, when the quantity of acid or alkali varies within moderate limits ?

(4) Does concentration of the solution affect the stability of a permanganate solution ?

(5) Does concentration of the permanganate solution during the boiling process promote the oxidation of organic substances ?

(6) If the answers to these questions are favorable to the use of a temperature of 100° C. for the oxygen process, what is the best way of carrying out the operation ?

(7) How do the results compare with those obtained when the ordinary temperature is employed, and what conclusions may be drawn from the working of the process ?

The points will be considered in the above order, and tables have been drawn up containing the experiments made to clear up these various points.

Table XI furnishes the answer to the first question by showing that a solution of permanganate of potash can be boiled in pure distilled water for one hour without suffering any reduction.

In all the experiments contained in the tables dealing with the various questions under consideration, a 24-oz. flat-bottomed glass flask was used, and the mouth of the flask simply covered by a wide glass funnel.

Table XII supplies the answer to the second question by showing that a permanganate solution containing 2·5 c.c. pure sulphuric acid can be boiled for two hours without suffering any reduction. After this length of time there is a slight reduction of permanganate, varying with the duration of boiling. This reduction seems to depend upon the concentration of the solution by loss from evaporation, as half a litre of water will stand boiling for two hours without any reduction of permanganate, whereas a quarter of a litre of water at the end of two hours shows a slight reduction of permanganate. Possibly the explanation of this is that, as the volume of water diminishes by evaporation, some of the permanganate adheres to the flask above the water-line, and is then reduced to a lower oxide of manganese.

Table XIII is not favorable to the use of alkaline solutions of permanganate for estimating the quantity of oxygen absorbed. The addition of caustic potash or soda caused a reduction of permanganate, even when the solution was only boiled for one hour. The quantity of oxygen used up in these blank experiments roughly varied with the quantity of caustic alkali used. This slight reduction of permanganate may be due to the caustic alkali in stick containing sulphur compounds, which act as reducing agents.

Table XIV furnishes the answer to the third question by showing that Edinburgh water practically absorbed the same quantity of oxygen, when the quantities of dilute sulphuric acid varied from 1 c.c. up to 20 c.c. It is evident that, when half a litre of water is used, there is no advantage in using large quantities of sulphuric acid.

The fourth question is answered by Table XII and the remarks made upon it. There is no reduction of permanganate during evaporation, if the concentration is kept within certain limits. A quarter of a litre of permanganate solution may be boiled for one hour, and half a litre for two hours, without suffering any reduction of permanganate, if certain precautions are observed. (See working details of the oxygen process at 100° C.) Roughly speaking, an acid solution of permanganate can be reduced to half its original volume by evaporation, whilst boiling, without suffering any reduction of permanganate.

Table XV answers the fifth question by showing that oxidation of the organic matter in Edinburgh water is not hastened by allowing a ¼ litre of the water to be boiled down to 60 c.c., if the loss of permanganate due to concentration—the quantity due to this cause being experimentally found—be deducted from the total quantity reduced.

The answers to these five questions are favorable to the use of a temperature of 100° C. for the oxygen process, if certain precautions are observed.

Working Details for the Oxygen Process at 100° *C.*

Experiments prove that a solution of permanganate and sulphuric acid can be boiled for two hours without any reduction of permanganate, if reducing agents are absent, and I shall now describe what I have found to be the best way of carrying out the process.

It is desirable to use, at least, half a litre of the sample of water, or if a smaller quantity must be used, it ought to be made up to half a litre with pure distilled water. For this purpose, organically pure distilled water is prepared from a good potable water containing excess of permanganate in solution. It is not necessary to use a litre of the sample of water, as the quantity is too bulky to manage readily, and the experimental error with half a litre of water should be so small as not to affect general results.

Solutions required :

Solution of permanganate of potash.

Dilute solution of pure sulphuric acid.

Solution of potassium iodide.

Solution of sodium hyposulphite.

Solution of starch.

The preparation of these solutions is given in Appendix C. Thoroughly cleanse a 24-oz., flat-bottomed flask with strong sulphuric acid, then rinse out well with tap water, and lastly with a small quantity of distilled water. Into this flask pour half a litre of the sample of water, add 10 c.c. of the permanganate solution and 10 c.c. of the dilute sulphuric acid; place a clean, moderately sized glass funnel into the neck of the flask, and stand the flask on a sheet of wire-gauze, covering the ring of an iron stand. It is advisable to put several small pieces of platinum-foil into the flask, so as to avoid bumping of the acid liquid. Bring the water to the boil, and then note the time; continue to boil gently for two hours over the wire-gauze, or transfer the flask to a sand-bath, where the water will boil gently without bumping, as the heat is thus more evenly applied to the flask. The more gently the liquid is boiled, the less will be the loss from evaporation—a point to be kept in view. It is best to use a glass funnel from which the stem has been filed off, so as to allow the watery vapour to escape easily; otherwise, the vapour will pass off at intervals with explosive violence, and thereby may cause partial loss of the liquid contents of the flask.

The water should be boiled for two hours, and, during this time, the loss from evaporation should not be more than a quarter, or a third, of the original volume; this will ensure that there is no reduction of permanganate due to a too great concentration of the liquid.

During the whole time the water is boiling, it should remain decidedly pink when the flask is held against the light ; if the pink colour becomes very faint, another 10 c.c. of the permanganate solution must be added.

At the end of two hours, remove the flask from the sand-bath and allow the water to cool for ten minutes ; then hasten the cooling by placing the flask in a basin of cold water, or by holding it under a stream of cold water running from the tap. When the water is reduced to the ordinary temperature, add 20 drops of the solution of potassium iodide for every 10 c.c. of the permanganate solution used. Agitate gently the contents of the flask, and the precipitate of manganic and manganese oxides will soon disappear, and give place to a clear, brilliant, yellow-brown colour. Run in the hyposulphite solution from a burette till the water is nearly colourless, then add enough of the solution of starch to render the solution a deep blue in colour, and continue to add the hyposulphite solution till the blue colour just disappears. The hyposulphite solution is standardised by adding 10 c.c. of the above potassium permanganate solution to ¼ litre of distilled water, also 10 c.c. of dilute sulphuric acid, and 20 drops of the solution of potassium iodide, and proceeding exactly in the way already described.

The quantity of oxygen absorbed by the organic matter is calculated in the same way as for the ordinary oxygen process, keeping in mind that 10 c.c. of the solution of permanganate are capable of yielding up 5 milligrms. of oxygen for oxidation purposes.

It is better to add sulphuric acid to the sample of water than to use permanganate alone, because sulphuric acid hastens the oxidation of organic matter.

For example, 20 milligrms. of wheaten starch, when boiled for two hours with sulphuric acid and permanganate solution, absorbed 17·28 milligrms. of oxygen; whereas, the same quantity, when boiled for the same time with permanganate solution without the addition of the acid, only absorbed 5·58 milligrms. of oxygen.

Irrespective of organic matter, the reducing agents likely to be present in a water and to reduce permanganate are—(1) chlorides, (2) nitrites, (3) ferrous salts, and (4) sulphuretted hydrogen.

The sulphuric acid will partially decompose the chlorides, forming hydrochloric acid. The quantity of hydrochloric acid formed in this way is very small, and becomes highly diluted in the ½ litre of water. Experimentally, it was found that a solution containing eight grains of chlorine per gallon practically reduced no permanganate through the formation of hydrochloric acid. Comparatively few waters contain so much chlorine, so the reduction of permanganate due to chlorides may be disregarded.

Nitrites and Ferrous Salts.—The oxygen absorbed, due to the presence of either of these compounds, may be estimated at the end of five minutes (see Appendix C).

Sulphuretted hydrogen, or nitrous acid, if present, can be boiled off after the addition of the sulphuric acid, before the permanganate is added.

The quantity of oxygen absorbed by the nitrites and ferrous salts must be deducted from the total quantity of oxygen absorbed, in order to find the oxygen absorbed by the organic matter.

TABLE XI

Experiments with a Solution of Permanganate of Potash in organically Pure Distilled Water.

Distilled water—was the distillate from Edinburgh City water, containing excess of permanganate in solution.

Solution of permanganate—10 c.c., capable of yielding up 5 milligrms. of oxygen.

Solution was kept at the boiling-point.

Capacity of flask was 24 oz. Glass funnel used.

Quantity of aqua destillata.	Quantity of permang. solution.	Duration of boiling.	Oxygen used.
I. ¼ litre	10 c.c.	1 hour	0·00 milligrm.
II. ¼ „	10 c.c.	„	0·00 „

TABLE XII

Experiments with Sulphuric Acid and a Solution of Permanganate of Potash in organically Pure Distilled Water.

In all cases 10 c.c. of a 1 in 4 solution of pure sulphuric acid were used.

The experiments were conducted under the same conditions as in Table XI. (See Strength of Solutions, &c.)

Quantity of aqua destillata.	Quantity of permang. solution.	Duration of boiling.	Oxygen absorbed.
I. ½ litre	10 c.c.	1 hour	0·00 milligrm.
II. ½ ,,	20 c.c.	1 ,,	0·00 ,,
III. ¼ ,,	10 c.c.	1 ,,	0·00 ,,
IV. ¼ ,,	20 c.c.	2 hours	0·00 ,,
V. ¼ ,,	10 c.c.	2 ,,	0·02 ,,
VI. ¼ ,,	10 c.c.	2 ,,	0·22 ,,
VII. ¼ ,,	10 c.c.	2¼ ,,	0·19 ,,
VIII. ½ ,,	10 c.c.	3 ,,	0·22 ,,
IX. ½ ,,	20 c.c.	3 ,,	0·12 ,,
X. ¼ ,,	10 c.c.	3 ,,	0·40 ,,
XI. ½ ,,	20 c.c.	5 ,,	0·31 ,,

Table XIII

Experiments with an Alkaline Solution of Permanganate of Potash in organically Pure Distilled Water.

Strength of permanganate solution, &c., the same as in Table XI.

In each experiment 10 c.c. of permanganate solution were used.   •

The caustic alkali used was in stick.

Quantity of aqua destillata.	Quantity of alkali.	Duration of boiling.	Oxygen absorbed.
I. 400 c.c.	1 grm. caustic potash	1 hour	0·12 milligrm.
II. 500 c.c.	2 grms. „	1 „	0·28 „
III. „	5 „ „	1 „	0·45 „
IV. „	4 „ caustic soda	1 „	0·36 „

Note.—In Experiments II and III *ordinary* distilled water was used, and 0·228 milligrm. of oxygen was deducted from the quantity of oxygen absorbed, as this was the quantity of oxygen absorbed by the distilled water containing sulphuric acid and permanganate.

TABLE XIV

Experiments showing that the Oxidising Power of a Solution of Permanganate is not affected by Variations of the Acidity of the Solution within Moderate Limits.

Conditions of experiments the same as in Table XI. In each experiment the duration of boiling was one hour.

Date.	Quantity of water.	Quantity of acid.	Quantity of permang. solution.	Oxygen absorbed.
1888 Nov. 20	½ litre Edinb. water	1 c.c.	10 c.c.	2·91 milligrms.
„ 20	„ „	2 c.c.	„	2·88 „
„ 21	„ „	2 c.c.	„	2·91 „
„ 21	„ „	5 c.c.	„	2·99 „
„ 22	„ „	5 c.c.	„	2·87 „
Dec. 15 ½	„ „	5 c.c.	„	1·60 „
„ 15	„ „	20 c.c.	„	1·63 „

TABLE XV

Experiments showing that Concentration of an Organic Solution does not hasten Oxidation in the Presence of Sulphuric Acid and Permanganate.

For these experiments it was convenient to use Edinburgh City water, because it is a peaty water, and, beyond a certain point, its organic matter is only slowly oxidised in the heat by acid and permanganate. Possibly the cellulose in peaty waters is only slightly oxidised by permanganate.

Strength of solutions the same as in Table XI. For every quarter-litre of water used 10 c.c. of permanganate solution were added, therefore when half a litre was used 20 c.c. of permanganate solution were added.

Date.	Quantity of water.	Volume reduced to	Duration of boiling.	Oxygen absorbed.	Oxygen used per ¼ litre.
1888. Dec. 20	¼ litre aq. dest.	60 c.c.	1 hour	Millig. 0·28	—
,, 20	,, Edin.water	,,	1¼ hours	1·73	(1·73—0·28) = 1·45 millig.
,, 18	½ litre ,,	Scarcely reduced	1 hour	2·85	1·42 ,,
,, 19	,, ,,	,,	1 ,,	3·02	1·51 ,,

Note.—Distilled water prepared as in Table XI

does not reduce permanganate, therefore 0·28 milli-grm. oxygen absorbed in the above experiment is due to concentration, and must be deducted from the oxygen absorbed by Edinburgh water in the above experiment.

TABLE XVI

*Table showing the Quantities of Oxygen absorbed by the
same Organic Solutions at Ordinary Temperatures,
and at 100° C.*

The quantity of the solution was made up to half
a litre with organically pure distilled water.

In all experiments 10 c.c. of a 1 in 4 solution of pure
sulphuric acid were used.

Excess of permanganate solution was added in all
cases.

The oxygen is stated as so much per cent.

Substance, 100 milligrms.	Ordinary temp. Oxygen absorbed in		Temp.=100° C. Oxygen absorbed in		Oxygen required (calculated).
	¼ hour.	3 hours.	1 hour.	2 hours.	
Cane sugar	0·0	0·0	97·4	111·2	111·9
Starch	0·0	0·2	—	86·4	118·5
I. Albuminoid solution from potato	1·9	6·1	31·6	—	32·1*
II. Ditto	—	6·2	—	67·8	67·3*
Albuminoid solution from wheaten flour	4·4	5·9	63·2	88·8	143·2*
Wheaten flour	0·5	1·0	70·3	84·2	—
Moist white of egg............	0·5	0·7	6·7	8·2	17·9*
Strychnia	9·2	14·1	214·0	—	—
Brucia	26·5	48·9	184·9	—	—
Muriate of morphia	37·7	43·2	173·6	—	—
Fresh urine....................	0·1	0·2	1·1	—	—
Ditto	—	—	0·9	1·1	—

* Total solids are reckoned as albumen—only approximately correct—and the nitrogen was found by the sulphuric acid process.

TABLE XVII

Table showing the Quantities of Oxygen absorbed by Natural Waters at Ordinary Temperatures, and at 100° C.

The two oxygen processes were carried out in the way already described.

The oxygen is stated as parts per million.

Water.	Ordinary temp.		Temp. =100° C.	Ratio of oxygen in 3 hours to oxygen in 2 hours.
	Oxygen absorbed in		Oxygen absorbed in 2 hours.	
	¼ hour.	3 hours.		
Edinburgh City supply	0·64	1·62	6·13	3·8
„ „	0·64	1·46	5·72	3·9
I. From Peterhead	0·97	2·24	9·12	4·0
II. „ „	0·35	0·86	4·28	4·9
I. Well, Kirkmichael	0·06	0·12	0·90	7·5
II. „ „	0·18	0·24	0·61	2·5
III. „ „	0·07	0·20	1·37	6·8
IV. „ „	0·15	0·30	2·31	7·7
V. „ „	0·00	0·25	1·54	6·1
VI. „ „	0·00	0·10	0·88	8·1
I. Spring, Patna	0·00	0·00	0·50	—
II. Well „	0·00	0·12	0·73	6·0
Sewage water	5·44	18·90	101·25	5·3
„ „	3·47	13·59	107·45	7·8
„ „ purified by alum	0·61	3·37	26·67	7·9
„ „ before filtration	3·89	15·13	108·05	7·2

Table XVIII

Table showing the Quantities of Oxygen absorbed by Waters at the Ordinary Temperature for Three Hours, and at 100° C.

Oxygen is stated as parts per million.

Water.	Oxygen absorbed in ¼ hour.	Oxygen absorbed in 3 hours.	Oxygen absorbed in 1 hour.	Ratio of oxygen in 3 hours to oxygen in 1 hour.
Edinburgh City supply	0·76	2·18	5·94	2·7
„ „	0·91	1·93	6·30	3·1
„ „	0·86	1·66	5·32	3·2
St Margaret's Well	0·04	0·13	0·99	7·6
St Anthony's Well	0·00	0·00	0·20	—
Spring, Braids	0·02	0·07	0·91	13·0
„ Blackford Hill	0·06	0·21	1·35	6·4
School Well, Straiton	0·00	0·12	1·43	12·0
Stream „	0·56	1·60	4·20	2·6
I. Well „	0·06	0·11	1·05	9·5
II. „ „	0·26	0·42	2·25	5·9
I. Prison, Peterhead	0·70	1·86	8·66	4·6
II. „ „	0·92	1·80	7·42	4·1
III. Well „	0·56	1·24	4·96	4·0
IV. „ „	0·00	0·33	5·69	—
Spring, Edinburgh	0·00	0·00	0·00	—
Not used as drinking waters. { St Margaret's Loch ...	0·94	1·96	10·42	8·3
„ „ ...	0·88	1·43	10·28	7·2
Duddington „ ...	1·23	2·70	10·49	3·8
„ „ ...	1·26	2·98	11·12	3·8
Dunsappy „ ...	0·80	1·57	6·76	4·2
Water of Leith.........	2·43	7·16	32·10	4·4
„	2·29	6·92	32·80	4·7
Stream Liberton	0·87	1·71	5·90	3·4
„ Braids	0·37	0·75	3·60	4·8

TABLE XIX

Table showing that on boiling an Organic Solution containing Sulphuric Acid and Permanganate, Oxidation is very Rapid during the First Quarter of an Hour and much more Gradual afterwards.

Oxygen is stated as parts per million.

Date.	Water.	Oxygen absorbed in				
		¼ hour.	½ hour.	1 hour.	2 hours.	3 hours.
1888						
Nov. 22	Edinburgh City supply	5·16	—	5·94	—	—
„ 24	„ „	4·54	4·88	5·36	—	—
„ 26	„ „	—	—	5·74	5·98	—
„ 27	„ „	—	—	—	—	6·04
„ 29	„ „	—	—	7·02	—	7·96
1889						
Apr. 20	Dhu Loch, Peaty	—	—	27·10	28·21	—
„ 8	I. Well, Peterhead	—	—	7·56	9·12	—
„ 8	II. „ „	—	—	3·35	4·28	—

Comparison between the Results obtained by the Oxygen Process at Ordinary Temperatures, and at 100° C.

I. *Made-up Organic Solutions* (see Table XVI).—At ordinary temperatures, cane sugar practically absorbed no oxygen from permanganate during the first three hours, whereas, at 100° C. for one hour 10 milligrammes of cane sugar absorbed 9·74 milligrms. oxygen, and at the end of two hours the sugar was practically completely oxidised.

A solution of starch, when kept at 100° C. for two hours, absorbs upwards of 400 times more oxygen than during three hours in the cold.

Albuminoid solution from potato absorbed five times as much oxygen, and from wheaten flour sixteen times as much oxygen in the heat as during three hours in the cold.

Urine absorbed five times as much oxygen at 100° C. for two hours as in the cold for three hours.

Vegetable alkaloids are practically completely oxidised when kept at 100° C. for two hours, in some cases at the end of one hour, whereas in the cold not more than a quarter of the required oxygen is absorbed in three hours.

Conclusions from Table XVI.—(1) Some organic solutions, such as sugar and vegetable alkaloids, are completely oxidised if boiled for two hours with sulphuric acid and permanganate.

(2) Albumens are only partially oxidised by boiling in this way for two hours, the ratio between oxygen absorbed and oxygen required for complete oxidation seeming to vary with the state the albuminoid body is in. For example, the albumen from potato was completely oxidised at the end of two hours, whereas the solution of moist white of egg only absorbed about seven-sixteenths of the oxygen required to oxidise the albumen only of the moist white of egg.

(3) Starch, if boiled for two hours with acid and permanganate, absorbs rather less than three quarters of the oxygen required for complete oxidation.

II. *Natural Waters* (see Table XVII).—This table shows that the oxygen absorbed by waters, when boiled for two hours with sulphuric acid and permanganate, exceeds the quantity absorbed during three hours in the cold, in a ratio ranging from 2·5 to 1 up to 8 to 1 in the case of potable waters and polluted waters.

Oxidation of organic matter is very rapid as soon as the organic solution is brought to the boil, and after the first quarter of an hour the oxidation is much more gradual.

Classification of Waters for the Oxygen Process at
100° C. for two Hours.

A water is of great purity if it absorbs less oxygen than 2 parts per million.

It is of medium purity if the oxygen absorbed is between 2 and 4 parts per million.

A water is suspicious, unless peaty, if the oxygen absorbed is between 4 and 6 parts per million.

A water is not to be recommended, unless peaty, if the oxygen absorbed is over 6 parts per million.

In any case, the oxygen absorbed should not be much over 6 parts per million.

PART III

THE SULPHURIC ACID PROCESS FOR ESTIMATING
THE ORGANIC NITROGEN

THE chief objections to the Albuminoid Ammonia
Process are—

(1) The ammonia obtained by this process does not
represent the whole of the nitrogen of albuminoid
bodies. Mr Wanklyn states that albumen yields about
two-thirds of its nitrogen as ammonia to alkaline permanganate, and the analyses contained in the following tables bear out this statement. Albuminoid
bodies play an important part in the organic impurities of waters, being present whether the organic
matter is of animal or vegetable origin.

It will be shown later on that urine, even when
highly diluted, only yields half of its nitrogen as ammonia to alkaline permanganate.

Vegetable alkaloids yield varying proportions of
their nitrogen as ammonia by Mr Wanklyn's process.

Although the same water may yield a fairly constant
proportion of its nitrogen as ammonia to alkaline permanganate, when analysed from time to time, yet two
waters taken from different sources do not necessarily
yield up their nitrogen in quantities having the same
ratio to the organic matters contained in these waters.
This is due to the ratio of nitrogen yielded as ammonia
to total nitrogen present being different for different
organic substances, or different for the same organic
bodies existing under different conditions. For example, the albuminoid solution from potato yielded up
about nine-tenths of its nitrogen as ammonia to alkaline permanganate, whereas an albuminoid solution
from wheat-flour only yielded up about three quarters
of its nitrogen as ammonia.

(2) Occasionally, waters are met with; which yield
up the nitrogen as ammonia so slowly that it is difficult to exhaust the water of its albuminoid ammonia.
This point may be of no importance in determining
the fitness, or unfitness, of the water for domestic purposes, but it imparts a degree of incompleteness to the
operation.

(3) Solutions of alkaline permanganate, even when
carefully prepared, contain traces of ammonia; and this
has to be allowed for in each analysis, or the solution
of alkaline permanganate must be boiled free of ammonia, just previous to use. The caustic potash seems
to contain a trace of organic matter, which is gradu-

ally acted upon by the alkaline permanganate, giving rise to ammonia.

The process for estimating the organic nitrogen, brought out by Drs Frankland and Armstrong, has failed to satisfy analysts. Its allegèd inaccuracy, the delicacy of apparatus required, and the necessity of gas-analysis, have prevented this process from coming into general use.

Kjeldahl in 1883 published a paper proving that, if organic substances containing nitrogen are charred by strong sulphuric acid in the heat, all the nitrogen is converted into sulphate of ammonia, unless the nitrogen is present as azo-, diazo-, or nitro-compounds. He also proved that the results obtained are as accurate as by the " Will and Varrentrapp Process." Kjeldahl applied the process to solid or semi-solid organic compounds, and used oxalic acid to estimate the quantity of ammonia obtained.

Is the Sulphuric Acid Process applicable to the Analysis of the Organic Matters contained in Potable Waters ?

In order to answer this question the following points must be considered :

I. During the evaporating down of a liquid containing ammonia, is any of the ammonia lost if the liquid be strongly acidified by sulphuric acid previous to heating ?

II. If water containing both free and organic ammonia is evaporated down to a small bulk—say one fifth of its original volume—without adding any acid, will the free ammonia only be driven off, or will some of the organic ammonia be lost due to the heat decomposing part of the organic bodies present ?

In answer to the first question, it may be shown that a solution of a known quantity of ammonium chloride, when strongly acidified with pure sulphuric acid, can be subjected to the process of distillation without the distillate containing any ammonia. It may be also proved that the ammonia is in the residual water left in the retort after distillation, if the solution be rendered alkaline by a solution of caustic soda, which has been boiled free of all ammonia, and distillation resumed. It is found that the distillate now contains all the ammonia originally added, as well as any ammonia the sulphuric acid may have contained.

The experiments with natural waters (see Table XX) give the answer to both questions. In this table it is shown that if a water is selected which contains both free and organic ammonia; if the free ammonia is measured by taking ½ or 1 litre of the water and distilling off all the free ammonia, and nesslerising the distillate ; if the total ammonia—free and organic

—is found by subjecting a $\frac{1}{4}$ litre of the water to the sulphuric acid process; if the organic ammonia is found by boiling down a $\frac{1}{4}$ litre of the water to 50 c.c., then adding 10 c.c. of pure sulphuric acid and conducting the sulphuric acid process to be immediately described; then it is found that the total ammonia obtained is equal to the sum of the free and organic ammonia.

These experiments prove—

(1) That an organic solution, when rendered strongly acid by sulphuric acid, does not lose any of its contained ammonia during the evaporation of the water, and the charring by sulphuric acid.

(2) That a neutral organic solution can be evaporated down to, at least, one-fifth of its original volume without any loss of organic ammonia.

The waters experimented with in Table XX are of various degrees of purity, ranging from the pure well water from Patna down to the Water of Leith water at Bonnington, which is little better than sewage water. It should be borne in mind that the ammonia was measured in the usual way by Nessler's reagent, and that by this plan the experimental error may be $\frac{1}{1000}$ of a milligrm. of ammonia for each 50 c.c. of distillate nesslerised, and that when $\frac{1}{4}$ litre of water is used the total error must be multiplied by four if the ammonia is stated as parts per million. Bearing these facts in mind, it is seen how closely the

three operations for the free, organic, and total
ammonia tally with one another.   It follows that the
sulphuric acid process for organic nitrogen is applicable to potable waters.

*Working Details for the Sulphuric Acid Process for
Organic Nitrogen.*

Apparatus and reagents necessary :—

(*a*) A glass-stoppered retort of ¾ litre capacity.

(*b*) A distillation flask of ½ litre capacity.

(*c*) A Liebig's condenser and stand.   Also an iron
stand with rings.

(*d*) A ¼ litre measuring flask ; a 10 c.c. and a 2 c.c.
pipette ; a 50 c.c. burette, with Geissler's stopcock,
graduated to $\frac{1}{10}$ of a c.c. ; several nesslerising
glasses, each marked at 50 c.c. ; several glass-stoppered flasks of 50 c.c. capacity, with the 50 c.c. mark
in neck of flask ; also a 100 c.c. glass-stoppered flask.

(*e*) Nessler's reagent.

(*f*) A weak solution of ammonium chloride.

(*g*) Distilled water, 50 c.c. of which should give no
colour reaction with Nessler's reagent.

(*h*) A solution of pure arsenious acid.

For preparation of these reagents see Appendix A.

(*i*) Pure sulphuric acid.

(*k*) Solution of caustic soda.

Preparation of Pure Sulphuric Acid.

The pure distilled acid can be bought, and, if 10 c.c. of the acid do not contain more than 0·02 milligrm. of ammonia, the acid may be used as it is, or subjected to the following process :

If the distilled acid cannot be obtained, the acid may be distilled in the laboratory from a glass-stoppered retort into an air-tight glass receiver. The acid may be prepared very pure in this way, but the process is slow; generally 10 c.c. of this distilled acid will contain 0·005 milligrm. of ammonia. An easier process for purifying the acid of its ammonia is based upon the fact that nitrous acid and ammonia, when heated together, mutually destroy one another, forming water and free nitrogen, thus : $NH_3 + HNO_2 = 2H_2O + N_2$.

Manufacturers of sulphuric acid, who use platinum retorts, are in the habit of adding sulphate of ammonia to the acid in order to destroy the nitrous acid in this way, as it injures the platinum retorts.

Mr Raphael Meldola and Dr E. R. Moritz[*] have drawn attention to the reverse process, namely, that by adding excess of potassium nitrite to sulphuric acid and heating for two hours and a quarter, the

[*] See number of 'Journal of Society of Chemical Industry' for February 29th, 1888.

ammonia is destroyed, unless the nitrogen is present as a nitro-, nitroso-, azo-, or diazo-compound.

I find the best way to obtain the acid free of ammonia is as follows :—Find out by experiment how much ammonia 5 c.c. of the sulphuric acid contain. This is done by neutralising and making alkaline the diluted acid with a solution of caustic soda boiled free of ammonia, and nesslerising the distillate. Next take 100 c.c. or ¼ litre of the sulphuric acid; put it into a clean glass-stoppered retort; add sufficient permanganate of potash to render the acid moderately green in colour, and heat on a sand-bath till the acid becomes again colourless. This ensures the complete oxidation of any organic compounds which may have got into the acid since it was manufactured. Calculate the quantity of potassium nitrite necessary to destroy all the ammonia contained in the acid, and add excess of nitrite to the acid; continue to heat the acid on a sand-bath. Collect the distilled acid as it drops from the end of the retort, and test, from time to time, for nitrous acid by collecting a few drops of the distillate, diluting it with water, and adding a drop of a solution of potassium iodide; add a few drops of starch solution, and if nitrous acid is present the solution strikes a blue colour. Continue to heat the sulphuric acid till there is no blue reaction obtained in this way from the distillate. This precaution ensures that the acid in the retort is now free of any nitrous acid.

Now take 10 c.c. of the acid in the retort; drop it into 30 c.c. of ammonia-free distilled water; pour it into a clean distillation flask, and render the solution alkaline with ammonia-free solution of caustic soda. On distilling and nesslerising the distillate there is generally no colour reaction obtained. If there is any ammonia still present the quantity is very small, and if the mean of three estimations of the ammonia is taken, the quantity of ammonia is a perfectly constant factor. The cold acid should now be run out of the retort into a perfectly clean ¼ litre bottle, having a well-fitting glass stopper.

Preparation of the Solution of Caustic Soda.

The solution, when ready for use, should contain neither ammonia nor nitrous acid. The commercial caustic soda in stick sometimes contains a good deal of nitrite. If a solution containing a small quantity of sulphate of ammonia is rendered alkaline with a solution of caustic soda containing sodium nitrite, and submitted to distillation, there is a slight loss of ammonia.

Caustic soda is better than caustic potash, as a solution of sodium sulphate can be concentrated further than one of potassium sulphate before it begins to bump.

If pure caustic soda cannot be obtained, a very pure solution of it may be prepared in the following way. Procure crystallised sodium carbonate, and if it contains nitrous acid, free it of this impurity by recrystallising the carbonate. In a litre of boiling water—good tap-water does if it is free of nitrates, nitrites, and contains little organic matter—dissolve 350 grms. of this sodium carbonate. Bring the solution to the boil in a large flask, add 75 grms. of slaked lime, freshly prepared from properly burned limestone; continue to boil the mixture on a sand-bath for ten minutes, then remove and let stand till the supernatant liquid is quite clear. The clear liquid is syphoned off into a glass-stoppered bottle. A solution so prepared contains no nitrites, very little ammonia, and practically no organic matter. The strength of the caustic solution is about 8 per cent. of sodium hydrate, and a $\frac{1}{4}$ litre of the solution will more than neutralise 10 c.c. of sulphuric acid. This solution, or one of similar strength prepared from caustic soda in stick, when about to be used, is boiled in a distillation flask or retort till 25 c.c. of the distillate give no colour reaction with Nessler's reagent.

The Sulphuric Acid Process.

If the retort has not been used for the same purpose before, it must be made clean by heating some sul-

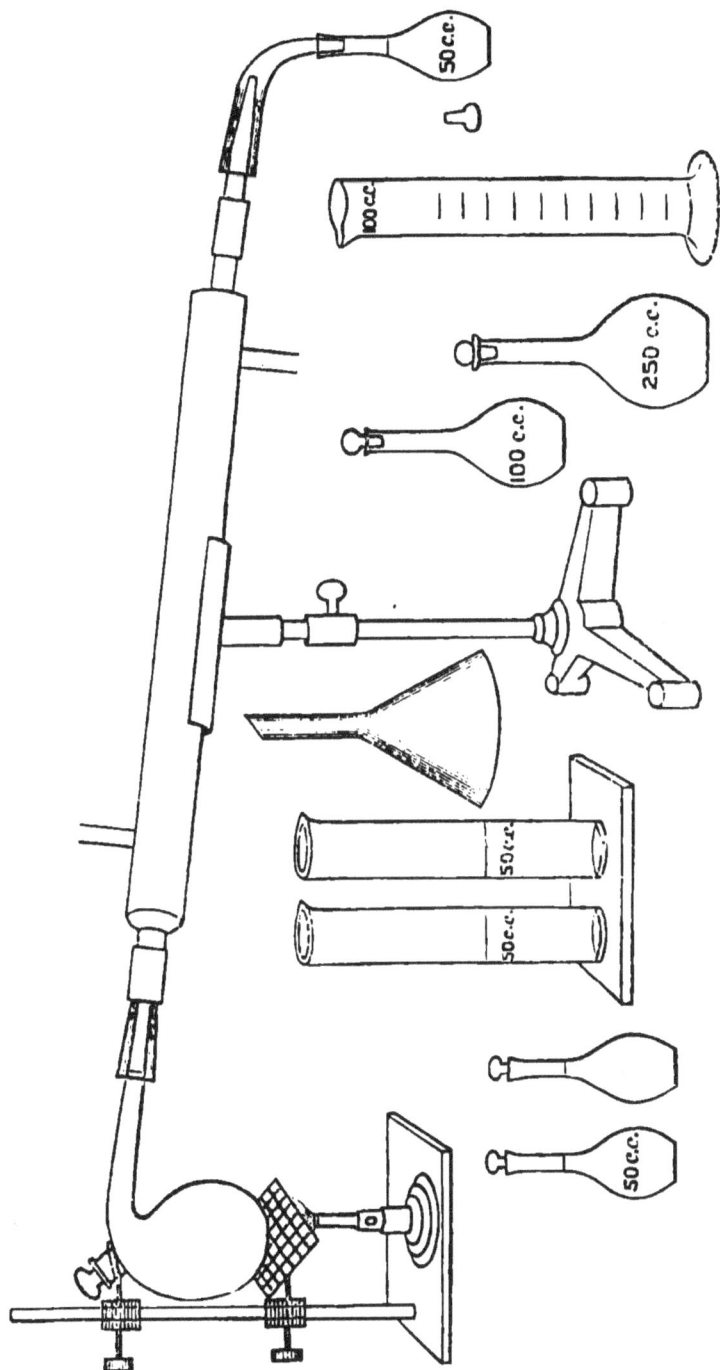

FIG. 1.

APPARATUS FOR THE SULPHURIC ACID AND ALBUMINOID AMMONIA PROCESSES.

5

phuric acid in it for some time, so as to char off all traces of organic matter. The retort is afterwards well washed out with tap water till the drippings do not redden blue litmus paper. Connect the retort with a Liebig's condenser, in the usual way, supporting the retort on a piece of wire-gauze covering the ring of an iron stand. The retort is fixed securely by passing another ring of the iron stand over the tubulated part of the retort. Measure out a ¼ litre of the sample of water, and pour it into the retort through a glass funnel. Replace the stopper of the retort, and adjust the condenser so that the inner third of the funnel-shaped end of the retort is at an angle to allow the water to run back into the retort; this arrangement prevents any loss of the water by bumping during the boiling process. If the water is acid, add a little freshly ignited sodium carbonate. Boil the water briskly till its bulk is reduced to about 50 c.c., then let it cool down. When nearly cold, add 10 c.c. of the sulphuric acid, which has been prepared in the way already described, also 10 c.c. of the solution of sodium arsenite. Replace the wire-gauze by a small sand-bath, and apply heat more gently. The remaining 60 c.c. of water will gradually distil off, and the distillate must be tested to see if it contain any nitrous acid; the most delicate test for nitrous acid is potassium iodide and starch. If nitrates are present in the water, the arsenious acid will reduce the nitric acid, converting it into lower oxides of nitrogen, which ultimately nearly all collect in the distillate as nitrous acid. If the distillate contain no nitrous acid, the water in the retort is allowed to distil off uninterruptedly till the sulphuric acid begins to fume. If nitrous acid distils over, the water must not be allowed to concentrate sufficiently to allow the sulphuric acid to char, or some of the organic ammonia will be destroyed by the nitrous acid. To prevent this over-concentration, 30 c.c. of distilled water must be added when the contents of the retort are reduced to about 30 c.c. if, when this stage is reached, there is still nitrous acid distilling over. This addition of distilled water may require to be repeated if there is much nitric acid. Whenever a few drops of the distillate give no colour reaction with potassium iodide and starch, the concentration of the sulphuric acid is allowed to go on. Hard waters are certain to bump whilst the last 10 or 20 c.c. of water are distilling off, and the flame must be moderated sufficiently to prevent this bumping. If a few small pieces of platinum-foil have been put into the retort—the foil should be heated to redness before it is put into the retort—the bumping is less likely to occur. Whenever the acid is sufficiently concentrated to char, it will assume a brownish colour, if there is much organic matter in the water; this is shown well in peaty waters and sewages, as the quantity of organic carbon is considerable. Purer waters will scarcely tinge the sulphuric acid. Whenever the acid is sufficiently strong to fume freely the Liebig condenser is raised higher, and this allows the acid fumes to condense back into the retort from its funnel-shaped end. During the charring process the flame may be increased, as the liquid ceases to bump as soon as the acid begins to fume.

Continue to heat the acid till it again becomes colourless, and in cases where it has not been coloured at all, it is well to keep the acid fuming for two hours. Hence, in all cases, it is advisable to keep the acid fuming for two hours at least, and then allow it to cool.

When the acid is cold, carefully drop in 30 c.c. of ammonia-free distilled water, so as not to crack the retort by overheating it. Now disconnect the retort from the condenser, and flush the condenser well with tap water, in order to wash out any sulphuric acid on the inside tube. The acid contents of the retort may be now transferred to a distillation flask, or retained in the retort. If the former course is followed, the acid contents of the retort are carefully run into the distillation flask through a clean glass funnel; the retort is washed out with a little distilled water, and lastly several times with some of the *cold* solution of caustic soda, and the washings are poured into the distillation flask. Continue to run into the flask the caustic soda solution till, after gentle agitation, the contents of the flask are decidedly alkaline to litmus. If the retort is preferred, some of the solution of caustic soda should be run into the retort through its funnel-shaped end, so as to wash back into the retort the acid adhering to the glass, and more caustic solution is added till the liquid in the retort is alkaline. Connect the retort, or distillation flask, with the Liebig condenser, using tinfoil as packing material if necessary. Bring the contents of the retort to the boil, and distil, at a moderate rate, into the 100 c.c. glass-stoppered flask. The necessary quantity of caustic soda solution generally makes the contents of the retort up to about 250 c.c., and by the time 100 c.c. have been distilled off all the ammonia will be in the distillate, if distillation has not been too rapid.

The Nesslerising Process.—Have some clean nesslerising glasses ready standing on a white tile, and some of the weak solution of ammonium chloride in a burette. Shake up well the contents of the 100 c.c. flask containing the distillate, and then divide the distillate between two of the 50 c.c. stoppered flasks. By means of a pipette, remove 10 c.c. of the distillate from one of the flasks, and pour it into a nesslerising glass; add 2 c.c. of Nessler's reagent; if the ammonia present seems to be less than $\frac{1}{50}$ milligrm. ($=2$ c.c. of the solution of ammonium chloride), rapidly add the remaining contents of the flask from which the 10 c.c. were removed. Try to imitate the tint of colour obtained, by running into another nesslerising glass as much solution of ammonium chloride as you think necessary; make the volume up to 50 c.c. with ammonia-free distilled water, and then add 2 c.c. of Nessler's reagent. Gently mix the contents of the glass; if the yellow tint obtained is too light, run in more ammonium chloride solution till the shade of colour is the same as that of the distillate. If the shade of colour is too dark, a new portion must be tried in another nesslerising glass.

The 50 c.c. of distillate in the other flask is used as a control experiment to prove the correctness of the estimated quantity of ammonia in the first 50 c.c. If 10 c.c. of the distillate contain over $\frac{1}{50}$ milligrm. of ammonia, the quantity of ammonia in 10 or 20 c.c. of the distillate should be measured, and the quantity in 100 c.c. of distillate thus calculated. When the nesslerising glass contains more than $\frac{1}{10}$ milligrm. of ammonia, the difference in the shades of colour is not so easily gauged as with a smaller quantity of ammonia. It is important to have all the nesslerising glasses of the same calibre, so that the 50 c.c. marks are all on the same level, otherwise the same quantities of ammonia may give different shades of colour.

Free Ammonia.—If the condenser be washed out before commencing distillation, the first 100 c.c. of the 200 c.c. of distillate taken off from the quarter of a litre of water will contain all the free ammonia. Test if there is much ammonia by nesslerising 10 c.c. of the 100 c.c.; if the ammonia is small in quantity the remaining 90 c.c. can be nesslerised in one glass, correcting the quantity of ammonia obtained by allowing for the 10 c.c. thrown away. If there is much ammonia 20 or 50 c.c. of the distillate can be standardised against the solution of ammonium chloride.

The free ammonia and organic ammonia are expressed as parts per million, or milligrammes per litre, if the respective quantities obtained from the quarter-litre of water are multiplied by four.

The organic ammonia is expressed as organic nitrogen if the organic ammonia is multiplied by 0·824.

If the 10 c.c. of sulphuric acid used contained any ammonia, and the water contained no nitrates or nitrites, this quantity must be deducted from the organic ammonia obtained, in order to find the correct quantity of ammonia. If the 10 c.c. of acid only contained a trace of ammonia—say 0·005 to 0·01 of a milligramme—and the water contained nitrates, there is no occasion to allow for this ammonia, as the nitrous acid formed from the nitric acid will destroy this small quantity of ammonia.

There is no occasion to add powdered permanganate to the contents of the retort, during the charring process, as the quantity of organic matter in a water is so small that the sulphuric acid will completely oxidise it within a reasonable time. There is always some danger of introducing ammonia, if powdered permanganate is used.

Explanation of some of the Details of the Process.

The retort and Liebig's condenser are preferable to small flat-bottomed flasks, because by their 'use' external ammonia is excluded, whereas with small flasks ammonia is taken up from the gas flame and fixed by the acid.

The Use of Arsenious Acid.—This acid is superior to ferrous chloride as a reducing agent in the sulphuric acid process, because an alkaline solution of arsenious acid can be prepared free of ammonia, whereas ferrous chloride cannot be prepared in this way.

If a solution of ammonium chloride and potassium nitrate is heated for two hours with sulphuric acid, after all water is driven off, there is a partial loss of ammonia, due to part of the nitric acid being reduced to lower oxides of nitrogen and destroying some ammonia.

If a water containing nitric acid and reducing agents, such as ferrous salts, is operated upon by the sulphuric acid process, there is a partial or complete destruction of the organic ammonia, according to circumstances.

It is for these reasons that the arsenious acid is

added to the water when its bulk is reduced to 50 c.c., in order to destroy all the nitric acid, if present, before the sulphuric acid begins to char the organic matter.

Remarks on the Sulphuric Acid Process.—The experiments contained in the various tables prove that by this process it is possible to accurately measure the total nitrogen contained in albuminoid bodies, and other organic substances usually present in potable waters.

The process is not difficult to carry out, and if care be taken to always use clean apparatus and flush pipettes, &c., with tap water just previous to use, the sources of external ammonia are practically *nil*. It is well to keep the retort for this process only, and use it for nothing else. It is not advisable to distil, or nesslerise, in au atmosphere containing much ammonia.

TABLE XX

Table showing that on boiling down a Potable, or Polluted Water, to one-fifth of its Original Volume practically only Free Ammonia is driven off, and that no Organic Ammonia is lost.

Total ammonia obtained by adding sulphuric acid before heating the water.

Free ammonia estimated by nesslerising the distillate from $\frac{1}{2}$ litre of the water.

Organic ammonia obtained by adding sulphuric acid after boiling the water down to one-fifth of its original volume.

The ammonia is stated as parts per million.

Water.	Free ammonia.	By sulphuric acid process.		Free ammonia plus organic ammonia.
		Organic ammonia.	Total ammonia.	
Edinburgh water	0·015	0·290	0·310	0·305
I. Well, Peterhead	0·150	0·230	0·370	0·380
II. „ „	1·696	0·320	2·020	2·016
Water of Leith Water	1·040	3·690	4·690	4·730
I. Well, Patna	0·050	0·145	0·190	0·195
II. „ Kirkmichael	0·008	0·104	0·115	0·112
III. „ Peterhead	0·032	0·512	0·530	0·544
IV. „ „	0·024	0·320	0·330	0·344

TABLE XXI

Table showing the Quantities of Ammonia obtained from Solutions of known Substances by the Alkaline Permanganate and Sulphuric Acid Processes.

In the experiments with strychnia, brucia, and muriate of morphia the quantity of solution used only contained 1 milligrm. of the substance.

In the case of urea only $\frac{1}{10}$ milligrm. was used. Picric acid, 2 or 3 milligrms. were used.

The ammonia is stated as parts per milligramme of substance.

Substance.	Alk. permanganate Ammonia.	Sulphuric acid. Ammonia.	Total nitrogen in substance stated as ammonia.
Strychnia...............	0·070	0·113 } 0·103	0·101
,, ...	—	0·094 }	
Brucia	0·058	0·090 } 0·085	0·086
,, ...	—	0·080 }	
Muriate of morphia...	0·035	0·046	0·045
Urea	0·04	0·56	0·56
Picric acid	—	0·114 ⎫	
,, ...	—	0·122 ⎪ 0·1107	0·222
,, ...	—	0·096 ⎬	
,, ...	—	0·111 ⎭	

TABLE XXII

Showing the Quantities of Ammonia obtained from Solutions of unknown Quantities of Albuminoid Substances by the Alkaline Permanganate and Sulphuric Acid Processes.

The ammonia is stated as parts per 100.

Substance.	Ammonia by alk. permang.	Ammonia by sulphuric acid.	
Urine	0·99	1·90	
I. Solution of albumen from potato	2·97	3·38	
II. „ „ „	5·88	7·00	
„ „ flour	12·37	15·90	
„ „ „	—	15·15	15·73
„ „ „	—	16·15	
Solution of flour	1·20	1·84	
Moist white of egg	1·29	1·90	
„ „	—	1·90	1·90

TABLE XXIII

Showing the Quantities of Ammonia obtained from Ordinary Waters by the Alkaline Permanganate and Sulphuric Acid Processes.

The ammonia is stated as parts per million.

Water.	Ammonia by alk. perm.	Ammonia by sulphuric acid.
I. Edinburgh City supply ...	0·126	0·188 = 0·155 nitrogen.
II. „ „ ...	0·132	0·296 = 0·245 „
Well No. I. Peterhead	0·170	0·670 = 0·552 „
„ 2nd sample	0·152	0·550 = 0·453 „
Well No. II. Peterhead	0·100	0·230 = 0·189 „
„ III. „	0·178	0·320 = 0·263 „
„ IV. „	0·194	0·512 = 0·422 „
„ V. „	0·120	0·320 = 0·263 „
„ I. Kirkmichael......	0·037	0·070 = 0·057 „
„ II. „	0·037	0·044 = 0·036 „
„ III. „	0·040	0·104 = 0·857 „
„ IV. „	0·060	0·050 = 0·041 „
„ I. Patna	0·035	0·145 = 0·119 „
„ II. „	0·025	0·050 = 0·041 „
Water I. Eccles, Maidstone...	0·060	0·328 = 0·270 „
„ I. „ „ ...	0·116	0·300 = 0·247 „
„ II. „ „ ...	0·400	0 500 = 0·412 „
„ II. „ „ ...	0·390	0·510 = 0·420 „
„ III. „ „ ...	0·128	0·160 = 0·132 „
„ I. Maidstone	0·062	0·164 = 0·135 „
„ II. „	0·005	0·056 = 0·046 „
„ III. „	0·308	0·880 = 0·725 „
Water. Bearsted, Maidstone...	0·085	0·350 = 0·288 „

Table XXIV

Showing the Quantities of Ammonia obtained from Polluted Waters by the Alkaline Permanganate and Sulphuric Acid Processes.

The ammonia is stated as parts per million.

Water.	Ammonia by alk. perm.	Ammonia by sulphuric acid.
I. Water of Leith, Bonnington	2·72	3·75 = 3·090 nitrogen.
II. Ditto filtered through clayey soil	2·38	3·89 = 3·205 ,,
III. Sample II	2·06	2·29 = 1·186 ,,
IV. ,, III	2·28	3·69 = 3·040 ,,
V. III and IV mixed after precipitation by alum	0·80	0·98 = 0·807 ,,
VI. Duddingston Loch	0·28	0·99 = 0·815 ,,
VII. Dhu Loch, peaty water	0·29	0·72 = 0·593 ,,
VIII. Sewage water	99·00	190·00 = 156·560 ,,

TABLE XXV.

Showing the Quantities of Oxygen absorbed at the Ordinary Temperature and at 100° C., also the Quantities of Ammonia obtained by the Alkaline Permanganate and Sulphuric Acid Processes.

Oxygen and ammonia are stated as parts per million.

Water.	Temp. =ordinary. Oxygen in 3 hours.	Temp. =100° C. Oxygen in 2 hours.	Ammonia by alkaline permang.	Ammonia by sulphuric acid.
				Nitrog.
Well I. Kirkmichael ...	0·12	0·90	0·037	0·044 = 0·036
„ II. „ ...	0·24	0·60	0·037	0·070 = 0·057
„ III. „ ...	0·10	0·88	0·040	0.104 = 0·857
„ IV. „ ...	0·25	1·54	0·060	0·050 = 0·041
„ I. Patna	0·12	0·73	0·035	0·145 = 0·119
„ II. „	0·00	0·50	0·025	0·050 = 0·041
„ IV. Peterhead	2·24	9·12	0·194	0·512 = 0·422
„ V. „	0·86	4·28	0·120	0·320 = 0·263
I. Edinburgh water ...	1·62	6·13	0·126	0·188 = 0·155
II. „ „ ...	1·46	5·72	0·132	0·296 = 0·245
Dhu Loch, peaty	9·75	28·21	0·288	0·720 = 0·593

Remarks on the Tables dealing with the Sulphuric Acid Process.

The analyses of known substances show that all organic nitrogen is obtained as ammonia by this process, whereas the nitrogen obtained by the alkaline permanganate process fell short of the total ammonia, in all cases. Picric acid is said not to yield any ammonia to sulphuric acid, but in the sample analysed the average quantity obtained represented about half of the total nitrogen in the picric acid. It is an example of a nitro-compound, and in these cases sulphuric acid does not convert the nitrogen—or only a part of the nitrogen—into ammonia.

Table XXII shows that the nitrogen of albuminoid bodies is only partially converted into ammonia by alkaline permanganate, the quantity obtained being about two-thirds to three-fourths of the total nitrogen present. The sulphuric acid process was used to estimate the quantity of albuminoid bodies present.

Table XXV shows, at a glance, the very different results obtained by the oxygen process at ordinary temperatures and at 100° C., and by the alkaline permanganate and sulphuric acid processes.

A Classification of Waters based upon the Analyses of the Waters in Tables XXIII and XXIV by the Sulphuric Acid Process.

In waters of great organic purity the organic nitrogen is less than 0·06 part per million.

In ordinary potable waters the organic nitrogen is between 0·06 and 0·12 part per million.

Waters are suspicious (unless peaty) if the organic nitrogen is over 0·12 part per million.

Peaty waters are potable if the organic nitrogen is under 0·25 part per million.

Waters are condemned, even if peaty, if the organic nitrogen is over 0·32 part per million.

It is not advisable to lay down hard and fast rules, but these figures are approximately correct.

PART IV

THE SULPHURIC ACID AND PERMANGANATE PROCESS
FOR ESTIMATING THE ORGANIC CARBON

WHEN a watery solution of carbo-hydrates is boiled with sulphuric acid and permanganate there is complete, or almost complete, oxidation of these substances into carbonic acid and water. This is proved in the experiments dealing with the oxygen process at 100° C. In these experiments the quantity of oxygen required to convert sugar into carbonic acid and water was calculated, and it was found that the quantity of oxygen absorbed by this substance tallied, or nearly so, with the quantity calculated. It is evident that if the concentration of the sulphuric acid solution is pushed far enough the oxidising property of the acid and permanganate will be greatly increased. Sulphuric acid alone will oxidise carbon, if it is given time, but if aided by permanganate the oxidation process is very rapid. Organic bodies when heated with sulphuric

acid are charred, some of the carbon separating out
as such, some passing off as carbonic dioxide, carbonic
monoxide, or other volatile carbon compound. If an
oxidising agent—such as permanganate—is present
whilst this charring is going on, the carbon passes off
as carbonic acid, and no carbonic monoxide escapes
this oxidising action.   Oxalic acid is an illustration of
the conversion of carbonic monoxide into carbonic
dioxide when heated with sulphuric acid and permanganate.   If oxalic acid is heated with sulphuric acid
only, half of its carbon is converted into carbonic
monoxide, and half into carbonic dioxide.

The organic bodies present in ordinary waters are
completely disintegrated by sulphuric acid in the heat,
the nitrogen being converted into ammonia, the
hydrogen abstracted as water by the sulphuric acid,
and the carbon set free, or converted into volatile
carbon compounds.   Under these circumstances, it is
a matter of experiment to find out if all the carbon of
these organic bodies can be obtained as carbonic acid,
when an oxidising agent is used along with the sulphuric acid.

In order to do this, two things are necessary, namely:

(1) Reagents containing no carbon compounds.

(2) Efficient means for collecting and estimating
the carbonic acid.

Purity of Reagents.

Pure sulphuric acid, as it is sold, is generally free of carbon as such, and of volatile carbon compounds. To ensure that it contains no carbon compounds, capable of oxidation into carbonic acid, it is only necessary to heat some of the pure acid—say 100 c.c. —with a little powdered permanganate of potash till the acid again becomes colourless.

Permanganate of potash, when in solution, will in time purify itself of traces of organic matter, especially if the solution is acidulated with sulphuric acid. This oxidation process is effected rapidly by boiling the acidulated solution for some time.

The Collection and Estimation of the Carbonic Acid.

After trying various plans, all of which proved unsatisfactory, I concluded that the apparatus used ought to provide for—

(1) The retention of the carbonic acid formed in the apparatus containing the sulphuric acid, during the charring process.

(2) The complete expulsion of the carbonic acid from the apparatus, after the oxidation of the organic substance is finished.

(3) The exclusion of all extraneous carbonic acid from the commencement to the finish of the process.

(4) The estimation of the carbonic acid formed.

The apparatus suitable for these purposes is as follows :

(1) A retort (not tubulated) of 24 oz. capacity, of moderately thick and uniform glass.

(2) A receiver of 6 oz. capacity, which fits accurately to the retort, without allowing the retort to project into the bulb of the receiver.

The retort and receiver are held firmly together, when required, by a piece of stout india-rubber tubing, overlapping the line of junction sufficiently to render the joint air-tight to pressure from without.

The receiver is tubulated, and has fitted into the hole a piece of glass tubing, secured properly by stout rubber tubing overlapping the joint.

The piece of glass tubing is drawn out at its outer end, so that ⅛ inch rubber tubing will pass over it, and yet grasp the glass tubing firmly. Fit this end of the glass tube with a piece of rubber tubing about 5 inches long.

For the collection of the carbonic acid formed, a small flask of about 6 oz. capacity is required, fitted with a rubber stopper through which passes a piece of glass tubing, furnished at its outer end with a piece of ⅛ inch rubber tubing about 5 inches in length.

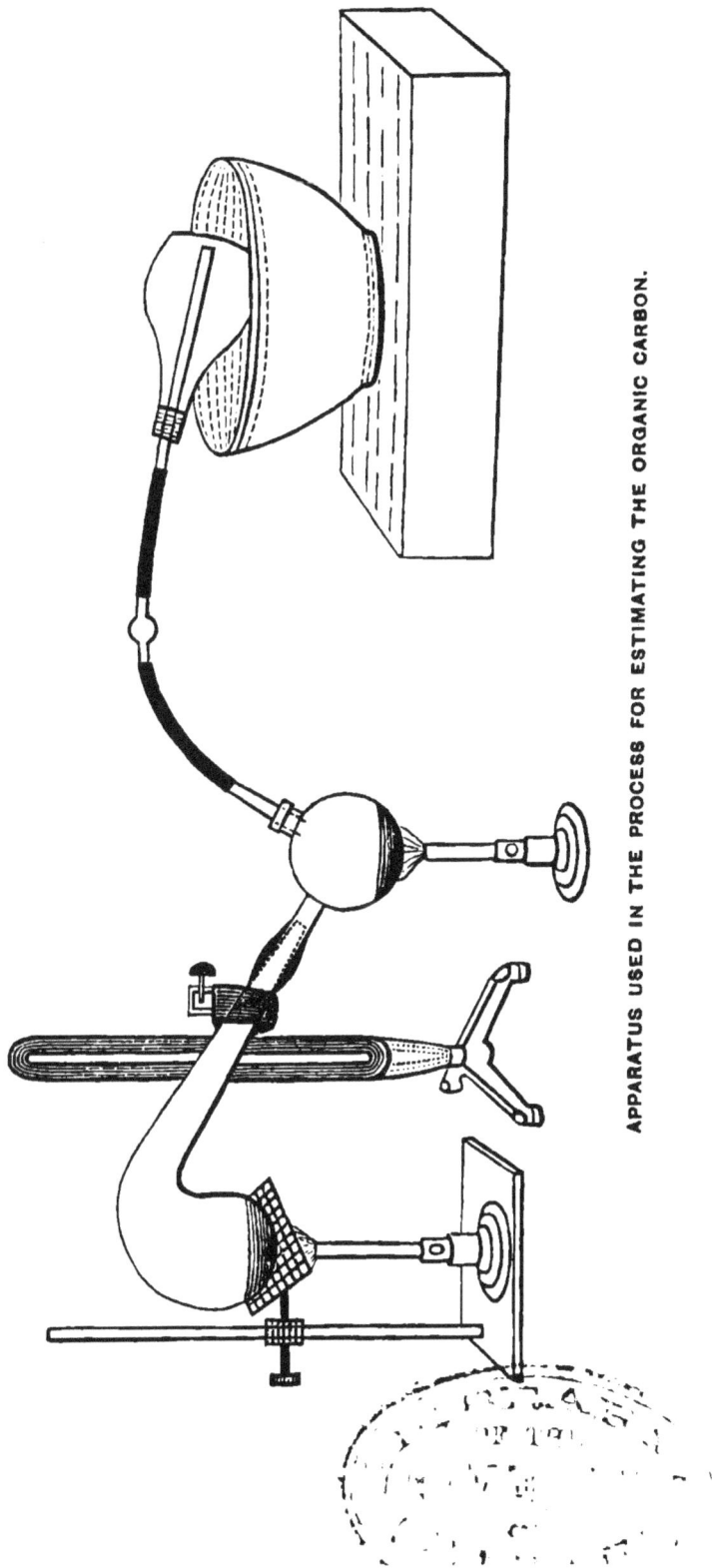

FIG. 2.

APPARATUS USED IN THE PROCESS FOR ESTIMATING THE ORGANIC CARBON.

Reagents.

The solutions required :

(1) Pure sulphuric acid.

(2) Solution of permanganate—strength 10 grms. to the litre.

(3) Solution of ferrous sulphate—strength such that 10 c.c. of the solution will decolorise 10 c.c. of the permanganate solution.

(4) Centinormal solution of barium hydrate.

(5) Centinormal solution of oxalic acid.

(6) Solution of phenol-phthalein.

For preparation of these solutions see Appendix E.

Working Details of the Process.

The retort is charred free of organic matter—unless used before—by heating some sulphuric acid in it. When purified in this way, wash it out well with tap water. Pour into the retort a ¼ litre of the sample of water; support the retort by means of a retort stand, and protect the bottom of the retort from the gas flame by means of a square of wire-gauze covering the ring of an iron stand. Briskly boil the water till its bulk is reduced to about 100 c.c., then add 2 c.c. of pure sulphuric acid, having previously diluted it ; continue to boil till the water is reduced to about 50 c.c. All carbonates, likely to be in the water, are decomposed on the addition of the acid, and the carbonic acid is boiled off by the time the volume is reduced to 50 c.c. At this stage, charge the clean receiver with 10 c.c. of the permanganate solution and 100 c.c. of distilled water; connect the receiver to the retort by folding the rubber tubing on the retort halfway back, passing up the receiver to the rubber tubing and then drawing the rubber tubing down over the upper end of the receiver. If the piece of rubber tubing is about 2½ inches long, and the connection properly made, this joint will be air-tight to any pressure exerted during the process. By means of a second Bunsen burner the permanganate solution is brought to the boil; the boiling of the water residue has been continued meanwhile; the permanganate solution in the receiver and the water residue are boiled for five minutes, ensuring a good flow of steam from the exit tube of the receiver. The boiling solutions fill the whole apparatus with steam, driving out the air and all carbonic acid in the retort and receiver. At the end of five minutes, suddenly clip the exit tube from the receiver, and remove both gas flames from under the apparatus. The clipping of the rubber tubing, attached to the glass tubing inserted into the receiver, may be accomplished by a metallic clip, or by simply kinking the rubber tubing down over the end of the glass tubing, and fixing it down by an elastic ring. If the joints are properly made, no air will enter the apparatus, but if it does so the fact is betrayed by the hissing noise made by the air as it leaks through the jointing. Whilst boiling the permanganate solution, previous to clipping the rubber tubing, it is important that no permanganate should bump into the water residue in the retort. This is prevented by fixing the apparatus in such a position that the retort is on a higher level than the receiver. Whenever the tubing is clipped, and the flames removed, tilt the whole apparatus so that the permanganate solution can flow gently into the retort. When the apparatus is in this position, it is easily noted if bubbles of air are oozing through the joint connecting the retort and receiver, as the air ascends through the permanganate in a stream of small bubbles. Previous to the addition of the permanganate solution, the volume of the water residue will be about 25 to 30 c.c., and when about 30 c.c. of permanganate solution have flowed into the retort do not allow any more to run in, but replace the apparatus in its former position, so that the receiver is once more on a lower level than the retort. Replace the wire-gauze by a small sand-bath—small concave plate of sheet iron covered with sand—and heat the contents of the retort very gently. In about an hour the 50 c.c. of watery vapour will have distilled into the receiver, leaving the acid and permanganate to act upon any organic matter

still unoxidised. After the sulphuric acid has fumed
for about half an hour all permanganate will have
been converted into manganous sulphate, and the acid
will be colourless. If it has a yellowish tinge it indi-
cates carbon still unoxidised—unless due to ferric
sulphate; and when the acid has cooled another 10
c.c. of the permanganate solution must be allowed to
flow into the retort.

When the sulphuric acid is colourless all carbon is
now present as carbonic acid.

At this stage, the probable gaseous contents of the
retort are carbonic acid, oxygen, chlorine (iodine or
bromine rarely), and nitric and sulphuric acid fumes.

When the acid is cold run gradually into the retort
the residual permanganate solution; agitate the contents of the retort, allowing the solution to alternately
flow from receiver to retort, and from retort to receiver.

The total liquid contents of the retort are about
150 c.c.; and on boiling these 150 c.c. what gases or
vapours will distil over? The small quantity of nitric
acid present, in any water, is too highly diluted by
150 c.c. of water to distil over, as a nitric acid solution must have attained a strength of about 8 per cent.
before acid vapours pass off. The distillate will then
contain only carbonic acid, oxygen, chlorine (from the
chlorides in the water), and watery vapour. Only the
carbonic acid and chlorine require to be taken into
consideration.

It is evident that if these two gases are distilled into a solution of barium hydrate both will unite with the baryta. The chlorine must be got rid of, as it will interfere with the estimation of the carbonic acid absorbed by the barium hydrate.

For this purpose, the contents of the apparatus are allowed to flow into the receiver, and then 10 c.c. of the ferrous sulphate solution are sucked up into a 10 c.c. pipette and passed into the receiver through the rubber tubing on the exit tube of the receiver, taking care that no air is allowed to enter into the receiver during this operation. This is secured by fixing the rubber tubing firmly on the end of the pipette, then unclipping the tubing and allowing the solution to flow in slowly, and by clipping the tubing again before the pipette is completely emptied of its contents. Ten c.c. of the ferrous sulphate solution will decolorise the permanganate and still remain in excess, as some of the latter has been destroyed already during the charring process. The contents of the receiver are now very gently heated—oscillating the liquid and the gas flame the whole time—as the liquid is being heated under diminished pressure, and will probably bump very violently unless this precaution is taken. When the liquid is heated to nearly the boiling-point, the whole apparatus is detached from its fixtures and taken into the hands, and the liquid allowed to flow backwards and forwards between the receiver and retort, shaking the contents well from time to time. This operation ensures that all the chlorine is absorbed by the liquid and acts on the ferrous sulphate, producing ferric salt and hydrochloric acid. At the end of ten minutes the reaction may be considered completed, and all chlorine converted into hydrochloric acid or ferrous chloride.

It only remains to boil off the carbonic acid into a solution of barium hydrate, and estimate the carbonic acid.

For this purpose the 6 oz. flask, fitted with the rubber tubing, is partly filled with distilled water—the bottom of the flask should be covered with about half an inch of water—and the water boiled till all air is expelled from the flask; then clip the rubber tubing in the usual way. When the flask has cooled run in 10 c.c. of the solution of barium hydrate, by means of a pipette, in the same manner as when the ferrous sulphate was passed into the receiver. It is essential to get the whole of the baryta solution into the flask, so the pipette must be allowed to empty itself till only a few drops of the solution are left in it; then clip the tubing and quickly withdraw the pipette from the tubing, when the remaining few drops of the solution will be sucked out of the pipette into the rubber tubing. Clip the end of the tubing with two fingers and loosen the clip, allowing the remaining baryta solution to flow into the flask. Restore the metallic clip to its former position, so as to prevent any air getting in. Distribute the contents of the retort so that half of the liquid is in the retort and half in the receiver; attach the small flask to the receiver by means of a connecting piece of glass tubing, with a bulb on it. The whole apparatus should be adjusted so as to rest properly on its supports, and the small flask is placed in a basin of cold water. Unclip the rubber tubing next the receiver, and proceed to boil the liquid in the retort. To prevent violent bumping this must be done gently, agitating the contents meanwhile, and using wire-gauze instead of the sand-bath. A few small pieces of platinum-foil, placed in the retort at the very beginning of the analytical process, will moderate this tendency to bumping. When the contents of the retort are boiling leave them to boil, and begin to boil the contents of the receiver in the same way; when nearly boiling unclip the tubing next the small flask, and adjust the apparatus so that there are no kinks in the rubber tubing.

The contents of the retort and receiver must be kept boiling briskly for about seven minutes, and the small flask kept cool by means of cold water. If the operation has been performed properly, there should be so little air in the apparatus that the rubber tubing next the small flask always remains partly collapsed. If the tubing becomes tense and hard some air

must have got in. The collapsed tubing shows that
there is no danger of any explosion, and it should be
the guide to indicate the state of pressure inside the
apparatus. After boiling for seven minutes, clip the
tubing attached to the small flask, and detach the
flask from its connections.

Briskly agitate the contents of the flask for five
minutes, and if the solution remains decidedly pink in
colour, outside air may be allowed to enter the flask
—it is best to allow the purest air which can be got
to enter the flask, such as air out of doors. If the
solution is faintly pink or colourless another 10 c.c.
of the baryta solution must be run in, and the contents
of the flask again well shaken.

From a graduated burette the centinormal solution of oxalic acid is run into the flask, till the pink
colour just disappears on the addition of one more
drop. Read off the number of c.c. of oxalic acid
solution required to neutralise the contents of the
flask.

Next standardise the baryta solution by taking 10
c.c. of the solution, and finding how much oxalic acid
solution is required to neutralise the 10 c.c. The
mean of three experiments should be taken as the
proper strength of the baryta solution. If the number
of c.c. of oxalic acid solution used by the contents of
the flask is deducted from the number of c.c. used
by 10 c.c. of baryta solution, the difference represents

the quantity of carbonic acid from the sample of water.

For example, 10 c.c. of baryta solution required 9·6 c.c. of oxalic acid solution, and 10 c.c. of baryta solution, plus the carbonic acid from a ¼ litre of water, used 1·5 c.c. of oxalic acid solution, therefore the difference between these two numbers is 8·1 c.c. of oxalic acid solution.

But 10 c.c. oxalic acid solution are equivalent to 4·40 milligrms. of carbonic acid, therefore 8·1 c.c. of oxalic acid $= \dfrac{4·4 \times 8·1}{10}$ milligrms. of carbonic acid $= 3·564$ milligrms.

Again, 4·4 milligrms. of carbonic acid $= 1·2$ milligrms. of carbon, therefore $\dfrac{3·564 \times 1·2}{4·4} = 0·972$ milligrm. of carbon.

When the answer is to be stated as carbon, the equation is more simple, thus :

10 c.c. oxalic acid solution $= 1·2$ milligrms. carbon, therefore $\dfrac{8·1 \times 1·2}{10} =$ carbon in the water stated as milligrammes.

If much carbonic acid is formed from the organic matter in the water there is a good deal of barium carbonate formed, and the solution, when neutralised with oxalic acid, will become pink again on standing. This is due to the fact that barium carbonate is alkaline in reaction, so the quantity of oxalic acid solution used at first to decolorise the solution must be taken as the proper quantity.

I have not had yet an extended experience with the application of this process for organic carbon to potable waters, but the results so far have proved satisfactory.

The solutions used to test the accuracy of the process were carefully prepared, and the results obtained show the method to be extremely accurate.

Very small quantities of the standard organic solutions were purposely used, as the object in view was the application of the process to water analysis. It is evident that by using decinormal or normal solutions of barium hydrate and oxalic acid, one would be enabled to use much larger quantities of organic substances, without impairing the accuracy of the process.

TABLE XXVI

Table giving the Results obtained when the Organic Carbon of Standard Solutions was estimated by the Sulphuric Acid and Permanganate Process.

All the solutions were made of the same strength, so that 1 c.c. of the solution contained 1 milligrm. of the pure organic substance.

Tartaric Acid.—Recrystallised and carefully dried.

Gallic Acid.—On drying it was found to contain 9·5 per cent. of water of hydration.

Salicine.—The pure substance used.

Sulphate of Quinine was carefully dried before use.

Substance.			Theoretical carbon.	Carbon obtained.	
Tartaric acid,	2	milligrms. used	0·640	0·678	Average for 2 milli-
,,	2	,,	0·640	0·624	grammes,
,,	3	,,	0·960	0·954	0·646.
Gallic ,,	2	,,	0·894	0·864	
,,	2	,,	0·894	0·924	0·884.
,,	2	,,	0·894	0·864	
Salicine	2	,,	1·091	0·092	Average for 2 milli-
,,	2	,,	1·091	0·996	grammes,
,,	2	,,	1·091	1·104	1·090.
,,	1½	,,	0·818	0·876	
Quin. sulph.	1	,,	0·643	0·642	
,, ,,	1½	,,	0·964	0·804	Average for 1½
,, ,,	1½	,,	0·964	0·912	milligrammes,
,, ,,	1½	,,	0·964	0·972	0·915.
,, ,,	1½	,,	0·964	0·924	

TABLE XXVII

Showing the Organic Carbon obtained from Natural Waters by the Sulphuric Acid and Permanganate Process.

Organic carbon is stated as parts per million.

	Organic carbon per litre.
Water Company, Maidstone	0·72
„ „ „	0·60
Well, near Maidstone, alleged to be polluted ...	4·95

APPENDICES

APPENDIX A

*Preparation of Solutions used in the Albuminoid
Ammonia Process.*

(1) *Nessler's Reagent.*—Dissolve 35 grms. of potassium iodide and 13 grms. of corrosive sublimate in 800 c.c. of distilled water, which has just been heated to boiling ; then add successive small quantities of a cold saturated solution of corrosive sublimate till the red iodide of mercury begins to be no longer dissolved, but remains as a permanent precipitate. Allow the solution to cool ; then render it strongly alkaline by adding 160 grms. of solid caustic potash, or 120 grms. of solid caustic soda, and make the solution up to one litre with distilled water. When the solution is cleared by settling, decant off the clear liquid and render the reagent sensitive by adding cautiously more cold saturated solution of corrosive sublimate till a permanent dirty yellow precipitate forms. Allow the reagent to settle in a well-stoppered bottle. When the liquid has again become clear some of the reagent is decanted into a smaller bottle

for use. Properly prepared Nessler's reagent has a yellowish tint, and at once strikes a yellow-brown colour with 5 c.c. of the weak ammonium chloride solution in 50 c.c. of water.

(2) *Stock Solution of Ammonium Chloride.*—Dissolve 3·15 grms. of pure ammonium chloride in crystals in one litre of distilled water. Each cubic centimetre of the solution contains 1 milligrm. of ammonia.

(3) *The Weak Solution of Ammonium Chloride.*— Take 10 c.c. of the above stock solution and make it up to one litre with distilled water. Each cubic centimetre of the solution contains $\frac{1}{100}$ milligrm. of ammonia.

This is the solution used for nesslerising.

(4) *Solution of Alkaline Permanganate.*—Dissolve 200 grms. of solid caustic potash and 8 grms. of pure permanganate of potash in one litre of distilled water; boil in an open flask till the volume is reduced to about 750 c.c., and make up to one litre with ammonia-free distilled water.

Fifty c.c. of this solution are used for half a litre of water in each analysis.

(5) *Carbonate of Soda.*—Solution made by boiling, for a short time, excess of the carbonate in water.

Ten c.c. of the solution are used, if the water to be analysed is acid in reaction.

(6) *Distilled Water.*—It must be free of ammonia. This is easily obtained by distilling an ordinary potable

water, which has been made slightly acid with a few drops of sulphuric or phosphoric acid. The acid prevents any ammonia from distilling over. A distillation flask or a still may be used.

These solutions (except No. 4) are also used for the sulphuric acid process.

(7) *Solution of Arsenious Acid.*—Used in the sulphuric acid process.

Dissolve by the aid of heat 50 grms. of the purest arsenious oxide in one litre of distilled water, which has been made moderately alkaline with caustic soda solution. The caustic soda solution must be free of nitrates and nitrites (see the Sulphuric Acid Process). When the arsenious acid is all dissolved the solution must be decidedly alkaline, or made so by adding more caustic soda solution. Boil down in a flask till the solution is reduced from 1000 c.c. to 900 c.c., and then make up to one litre with ammonia-free distilled water.

Ten c.c. of this solution are used for each operation.

APPENDIX B

Preparation of Solutions used in the Oxygen Process at Ordinary Temperatures.

(1) *Solution of Potassium Permanganate.*—Dissolve 0·395 grm. of the pure crystals in one litre of distilled water.

1 c.c. of this solution, with acid, yields 0·10 milligrm. of oxygen.

1 c.c. of this solution exactly oxidises 0·2875 milligrm. nitrous acid (NO_2).

1 c.c. of this solution exactly oxidises 0·2125 milligrm. hydrogen sulphide (H_2S).

1 c.c. of this solution exactly oxidises 0·9000 milligrm. ferrous oxide (FeO).

(2) *Dilute Sulphuric Acid.*—Add gradually 100 c.c. of the pure acid to 300 c.c. of distilled water. The basin containing the water should stand in cold water whilst adding the acid, in order to prevent a too rapid rise of temperature.

10 c.c. of this dilute acid are added to each sample of water examined.

(3) *Solution of Potassium Iodide.*—Dissolve 10 grms. of the purest iodide in 100 c.c. of distilled

water; five drops of this solution will decolorise more than 10 c.c. of the permanganate of potash solution.

(4) *Solution of Sodium Hyposulphite.* — Dissolve 0·775 grm. of the pure material in one litre of distilled water; about 40 c.c. of this solution will decolorise the iodine set free by 10 c.c. of the permanganate solution.

(5) *Solution of Starch.*—Add 1 grm. of powdered starch to 250 c.c. of boiling distilled water, boil for five minutes and filter. When the solution has settled syphon off the clear liquid into a stoppered bottle. A few drops of this solution will colour the quarter-litre of water of a deep blue if there is free iodine present. The solution gradually loses the power of producing this blue colour with iodine.

The Chemical Reactions which occur at the various stages of the Oxygen Process.

(1) Permanganate of potash in the presence of sulphuric acid is reduced by organic matter to potassium sulphate and manganous sulphate.

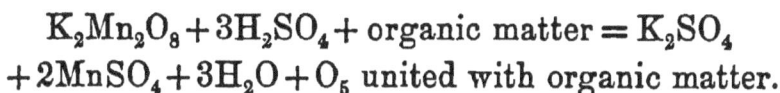

$$K_2Mn_2O_8 + 3H_2SO_4 + \text{organic matter} = K_2SO_4$$
$$+ 2MnSO_4 + 3H_2O + O_5 \text{ united with organic matter.}$$

(2) When potassium iodide is added to a solution of permanganate of potash in the presence of sulphuric

acid, potassium and manganous sulphates are formed and iodine is liberated.

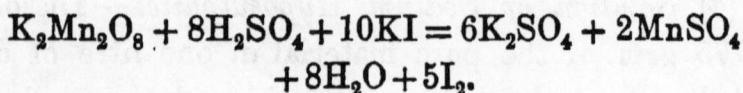

$$K_2Mn_2O_8 + 8H_2SO_4 + 10KI = 6K_2SO_4 + 2MnSO_4 + 8H_2O + 5I_2.$$

(3) The addition of sodium hyposulphite to the iodine solution decolorises the solution by producing sodium iodide and tetrathionate of soda.

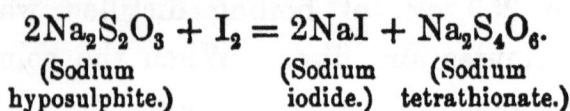

$$2Na_2S_2O_3 + I_2 = 2NaI + Na_2S_4O_6.$$
$$\text{(Sodium} \qquad \text{(Sodium} \quad \text{(Sodium}$$
$$\text{hyposulphite.)} \qquad \text{iodide.)} \quad \text{tetrathionate.)}$$

(4) A solution of permanganate of potash when boiled with hydrochloric acid is reduced to potassium and manganous chlorides with the liberation of free chlorine.

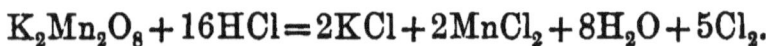

$$K_2Mn_2O_8 + 16HCl = 2KCl + 2MnCl_2 + 8H_2O + 5Cl_2.$$

A water containing much chlorides when boiled with permanganate and sulphuric acid will have its chlorides partially decomposed, and chlorine set free. The following reaction may be supposed to take place :

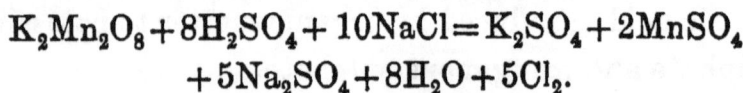

$$K_2Mn_2O_8 + 8H_2SO_4 + 10NaCl = K_2SO_4 + 2MnSO_4 + 5Na_2SO_4 + 8H_2O + 5Cl_2.$$

APPENDIX C

*Preparation of the Solutions used in the Oxygen
Process at 100° C.*

(1) *Solution of Permanganate of Potash.*—Dissolve
1·975 grms. of pure crystals in 1 litre of distilled
water.

10 c.c. of this solution can yield 5 milligrms. of
oxygen.

(2) *Dilute Sulphuric Acid.*—Add 100 c.c. of pure
acid to 900 c.c. of distilled water.

10 c.c. of this acid are used for each operation.

(3) *Solution of Potassium Iodide.*—Use a 10 per
cent. solution of the pure material.

Twenty drops will decolorise more than 10 c.c. of
the permanganate solution.

(4) *Solution of Sodium Hyposulphite.* — Dissolve
3·875 grms. of the pure material in 1 litre of distilled water. About 40 c.c. of this solution will decolorise the iodine set free by 10 c.c. of the permanganate solution.

(5) *Solution of Starch.*—Add 1 grm. of powdered
starch to 250 c.c. of boiling distilled water; boil
for five minutes and filter; let stand till settled, and
syphon off the clear liquid into a stoppered bottle.

(6) *Distilled Water.*—It is prepared free of organic matter from an ordinary potable water by adding excess of permanganate of potash, so that the colour of the water remains pink. Distil the water from a distillation flask or still through a condenser into a flask which fits closely to the condenser. Do not distil too far, or some permanganate may get into the distillate. If the contents of the distillation flask are made acid with sulphuric or phosphoric acid, the distillate will contain no ammonia.

APPENDIX D

The Estimation of Nitrates as Nitrous Acid by means of Sulphuric Acid and Arsenious Oxide.

Whilst carrying out experiments with arsenious acid in connection with the sulphuric acid process, the results obtained led me to believe that, under certain conditions, the whole of the nitric acid present in a sample of water could be collected in the distillate in the form of nitrous acid.

The conditions favorable to the total conversion of nitric into nitrous acid by sulphuric acid and arsenious oxide are—

(1) The acid solution must not be concentrated too much by distillation.

(2) The presence of air.

(3) Slow distillation.

(4) The quantity of nitric acid in the water must not be too large.

In the sulphuric acid process the first condition is present, as the concentration of the acid solution is not carried beyond a reduction of volume to 30 c.c., till all the nitric acid is reduced to lower oxides of nitrogen. This reducing action does not begin till the sulphuric acid has attained a strength of 15 to 20 per cent.

The second and third conditions are ensured by distilling very slowly, and condensing the watery vapour in a Liebig's condenser.

The fourth condition requires that the quantity of water used should not contain more than 6 milligrms. of nitric acid. If the nitric acid contained in the ¼ litre of water used for the sulphuric acid process is estimated by this method, as well as the organic nitrogen, then this is the outside quantity of nitric acid the water should contain. 6 milligrms. of nitric acid in ¼ litre of water are equivalent to 1·68 grs. of nitric acid per gallon.

The nitric acid is reduced to lower oxides of nitrogen, but in the presence of air the distillate ultimately contains all the nitrogen as nitrous acid. If any nitric oxide (NO) is formed, it takes up oxygen from the air and becomes nitrous acid (HNO_2).

If the nitric acid is large in quantity the nitric oxide is probably formed too freely to become properly oxidised, and some of it gets lost.

The process for collecting and estimating the nitrous acid formed is as follows:

During the sulphuric acid process, arsenious acid ($= 0.5$ grm.) and sulphuric acid ($= 10$ c.c.) are added to the ¼ litre of water when about 200 c.c. have been distilled off. At this stage put into a small flask, say of about 6 oz. in capacity, 1 c.c. of a caustic soda solution which contains no nitrites, add 10 c.c. of distilled water, and pass the lower end of the Liebig condenser into the neck of the flask so that the distillate drops into the solution. The condenser should be sloped at an angle, so that a drop of distillate always blocks the lower end of the inside tube of the condenser, and the upper end of the condenser should be properly attached to the retort, so that there is little chance of any gases escaping at this end of the condenser. Continue to distil, adding distilled water to the retort if necessary (see Sulphuric Acid Process), till a drop of the distillate gives no immediate blue reaction when acidulated with sulphuric acid, and a little potassium iodide and starch added.

Wash out the condenser with a little distilled water, and add the washings to the contents of the flask.

The small flask now contains all the nitric acid as sodium nitrite. It now remains to estimate the nitrous acid in the distillate. This is done by means of a standard solution of potassium nitrite, and a solution of metaphenylene-diamine.

The solution of metaphenylene-diamine is prepared by dissolving 0·50 grm. of the substance in 200 c.c. of distilled water, and adding sufficient sulphuric acid to render the solution acid.

The solution of potassium nitrite is prepared by dissolving 0·270 grm. of the purest material (or equivalent quantities of silver nitrite and potassium chloride) in one litre of water. Each c.c. of this solution is equivalent to 0·20 milligrm. of nitric acid. Prepare a 1 in 4 solution of sulphuric acid.

Dilute the distillate up to 250 c.c. with water and mix thoroughly. Measure out into a nesslerising glass 50 c.c. of the diluted distillate, add 2 c.c. of the dilute sulphuric acid and 2 c.c. of the solution of metaphenylene-diamine. Agitate gently the contents of the glass, and in a short time a yellow or reddish-yellow colour will appear, according to the quantity of nitrous acid present. The tint of colour must be imitated by running some of the standard solution of nitrite from a burette into another glass, diluting up to 50 c.c. with water and adding 2 c.c. of each of the above reagents. When an idea of the quantity of nitrous acid present is obtained, take a fresh 50 c.c. of the distillate, and the proper quantity of standard nitrite solution, and proceed as above. It is best to start the reaction in the solutions in the two glasses as near the same time as possible, and in about ten minutes the maximum tint of colour is reached. Fresh solutions must be made up till similar tints are obtained in both glasses. Fifty c.c. are the one fifth of the distillate, and one c.c. of the nitrite solution is equivalent to $\frac{1}{6}$ milligrm. of nitric acid; therefore the number of c.c. of nitrite solution which is equivalent to 50 c.c. of the diluted distillate represents so many milligrammes of nitric acid in the total distillate. If the number of c.c. used of the standard nitrite solution is multiplied by 0·222 the quantity of nitrogen (as HNO_3) in the water used is obtained.

If it is found that the nitric acid in the quarter-litre of water is above 6 milligrms., a smaller quantity of water must be used in order to get accurate results. In such a case, it is best to put 30 c.c. of distilled water into a retort, add 10 c.c. of sulphuric acid and 10 c.c. of arsenious acid solution, then put in 10 c.c. of the sample of water. Heat gently, and if only a light blue colour is produced by a drop of the distillate, when tested with the starch test, add other 40 c.c. of the water to the retort, and proceed as above. This method of estimation is exceedingly accurate with small quantities of nitric acid, and with a water containing much nitric acid, I have got perfectly accurate results when only using 10 c.c. of the sample of water in the above manner.

Experiments illustrating the Accuracy of the Method.

Potassium nitrate—4·0 milligrms.=0·554 milligrm. nitrogen.

First experiment, used 4 milligrms. and obtained 0·576 milligrm. nitrogen.

Second experiment, used 4 milligrms. and obtained 0·543 milligrm. nitrogen.

Third experiment, used 4 milligrms. and obtained 0·552 milligrm. nitrogen.

Potassium nitrate—10 milligrms.=1·386 milligrms. nitrogen.

First experiment, used 10 milligrms. and obtained 1·132 milligrms. nitrogen.

Second experiment, used 10 milligrms. and obtained 1·124 milligrms. nitrogen.

The last two experiments show that the method tends to be inaccurate (if carried out in the above way) when the nitric acid is present in too large a quantity.

The following series of waters had the nitrates and

nitrites estimated by the arsenious acid method, and by the aluminium process.

Where aluminium was employed, 50 c.c. of the sample of water and 40 c.c. of a 5 per cent. solution of caustic soda (free of nitrates and nitrites) were boiled down in an open flask to about 60 c.c.; then transferred to a tall glass cylinder (100 c.c. graduated measure), and when nearly cold two strips of clean sheet aluminium were added. The cylinder was plugged with an india-rubber cork having a glass tube passing through it containing glass-spun moistened with pure sulphuric acid. At the end of about twenty-two hours the contents of the cylinder and the glass-spun were transferred to a distillation flask, and the ammonia estimated in the usual way.

	Nitrogen stated as grains per gallon.	
	Arsenious acid.	Aluminium.
Water Company, Maidstone	0·31	0·29
Well, Ashford Road „	1·98	2·15
„ Paradise Road „	5·76	5·13
„ Bearsted „	1·08	1·01
„ Eccles „	4·27	4·11
Spout, Allington Lock „	0·76	0·70

In every water, except one, the arsenious acid process yielded more nitrogen than the aluminium process.

This is what may be expected, as there is no certainty about the aluminium having reduced all the nitric acid, unless the water is treated with aluminium for a second time; whereas with arsenious acid the absence of the blue reaction with starch shows when the distillate no longer contains nitrous acid. The above one exception is where the quantity of water used for the arsenious acid method contained 6·9 milligrms. of nitric acid.

APPENDIX E

Preparation of Solutions used in the Process for estimating the Organic Carbon.

(1) *Pure Sulphuric Acid.*—Add some powdered permanganate of potash to some pure sulphuric acid till the colour is moderately green, and then heat the acid in a flask, covered with a beaker, till it becomes colourless. 2 c.c. of the acid are used for each operation.

(2) *Solution of Permanganate.*—Dissolve 10 grms. of pure permanganate of potash in 1 litre of distilled water, add 1 c.c. of pure sulphuric acid; boil the solution for one hour, and keep in a glass-stoppered

bottle. 10 c.c. of the solution are used for each operation.

(3) *Solution of Ferrous Sulphate.*—Dissolve 50 grms. of the pure material in ½ litre of distilled water, add 1 c.c. of pure sulphuric acid, and boil for five minutes. Keep the solution in a glass-stoppered bottle. 10 c.c. of the solution will decolorise more than 10 c.c. of the permanganate solution.

(4) *Centinormal Solution of Barium Hydrate.*—Dissolve 3·15 grms. of crystallised hydrate in 1 litre of hot water, boiled free of carbonic acid. This solution is centinormal for dibasic acids. Make the solution strongly pink with phenol-phthalein.

(5) *Centinormal Solution of Oxalic Acid.*—Dissolve 1·26 grms. of the pure crystallised acid in 1 litre of distilled water. This solution is centinormal for di-acid bases. 10 c.c. of this solution will neutralise generally more than 10 c.c. of the baryta solution, because the baryta contains carbonate.

(6) *Solution of Phenol-phthalein.* — Dissolve 0·5 grm. of the substance in 100 c.c. of distilled water, aiding its solution by adding 5 c c. of rectified spirit. This solution is pink in alkaline, and colourless in acid or neutral solutions.

PRINTED BY ADLARD AND SON, BARTHOLOMEW CLOSE.

www.ingramcontent.com/pod-product-compliance
Lightning Source LLC
Chambersburg PA
CBHW021821190326
41518CB00007B/697